Energy Security and Cooperation in Eurasia

Why are bilateral relations, especially in the area of energy security, so different in the cases of U.S.–Russia, U.S.–Azerbaijan, and Russia–Germany energy deals? Why do some states find common ground despite differences, while others, with all the seemingly favourable conditions, are sinking into animosity?

Energy Security and Cooperation in Eurasia explores varying outcomes of energy cooperation, defined as diplomatic relations, bilateral trade, and investment in oil and natural gas. The book looks at economic potential, geopolitical rivalry, and domestic interest groups in the cases of U.S.–Russia, U.S.–Azerbaijan, and Russia–Germany energy ties. It looks at major projects in each case (Sakhalin and Arctic oil and gas production, Baku–Tbilisi–Ceyhan and Nord Stream pipelines) and activities of international oil companies. The book also provides a detailed analysis of the situation in Ukraine since 2014 and Russia's annexation of Crimea, and their effect on European energy security. This book utilizes an innovative approach of exploring the dyads of states (bilateral relations) along the economic, geopolitical, and domestic lobbying dimensions.

This book is a valuable resource for graduate and undergraduate students, academics, and researchers in the areas of security, political economy, comparative politics, and post-Soviet studies, as well as for the general public.

Ekaterina Svyatets is a lecturer at the University of Southern California, USA. She teaches economics and policy of natural resources and the environment.

Routledge Studies in Energy Policy

Our Energy Future
Socioeconomic Implications and Policy Options for Rural America
Edited by Don E. Albrecht

Energy Security and Natural Gas Markets in Europe
Lessons from the EU and the USA
Tim Boersma

International Energy Policy
The emerging contours
Edited by Lakshman Guruswamy

Climate Policy Integration into EU Energy Policy
Progress and prospects
Claire Dupont

Energy Security and Cooperation in Eurasia
Power, profits, and politics
Ekaterina Svyatets

Sustainable Urban Energy Policy
Heat and the city
Edited by David Hawkey and Janette Webb

For further details please visit the series page on the Routledge website:
http://www.routledge.com/books/series/RSIEP/

Energy Security and Cooperation in Eurasia

Power, profits and politics

Ekaterina Svyatets

First published 2016 by Routledge

2 Park Square, Milton Park, Abingdon, Oxfordshire OX14 4RN
52 Vanderbilt Avenue, New York, NY 10017

Routledge is an imprint of the Taylor & Francis Group, an informa business

First issued in paperback 2019

British Library Cataloguing-in-Publication Data
A catalogue record for this book is available from the British Library

Library of Congress Cataloging-in-Publication Data
A catalog record for this book has been requested

ISBN: 978-1-138-90261-9 (hbk)
ISBN: 978-0-367-33273-0 (pbk)

Typeset in Goudy Oldstyle Std
by Saxon Graphics Ltd, Derby

Contents

Acknowledgements

Writing this book, which evolved from my dissertation, has been a long, exciting, and challenging process. More than a final manuscript, it gave me a unique opportunity to learn and grow.

First and foremost, I am very grateful to my academic mentors – Professors Rob English, Pat James, Saori Katada, and Najmedin Meshkati. Their mentoring, kindness, friendship, and encouragement have enabled me to complete my book and to become a scholar. I thank them for their guidance and support, their countless hours with me, and generous help in every aspect of my student life. They made me feel at home at the University of Southern California (USC), and I have greatly enjoyed my graduate school years.

I also would like to thank my family in the United States – Nathan, Ann, Terry, and Zon Mundhenk, Luda and Olga Spilewsky, Indira Persad, Yulia Shmatkova, Masha Krugliakova, Victor Malychev, and Brett McBride. I have been very fortunate to have an amazing group of friends at USC who have been supportive of my professional growth – Adele di Ruocco, Christina Wagner Faegri, Maria Armoudian, Nadia Fomina, Mikhail Zibinsky, Jenifer Whitten-Woodring, Xiangfeng Yang, Jeffrey Fields, Meredith Shaw, Amy Below, Simon Radford, Nicolas De Zamaroczy, Jessica Liao, Parker Hevron, Alice Bardan, Zlatina Sandalska, Fabian Borges, Rebecca Kimitch and all my friends and colleagues. I am also grateful to Lena Latipova, Alena Egorina, Katinka Nedwed, Jack Matlock, Molly O'Neal, and Paul Almond for their friendship. I have also enjoyed the support of my friends and colleagues at the Environmental Studies Program at USC – Karla Heidelberg, Jim Haw, Lisa Collins, David Ginsburg, Rod McKenzie, Naomi Martinez, Jill Sohm, and Kristen Weiss.

I would like to express my special thanks to my mentor and friend Mary Ann Murphy of the USC American Language Institute. She helped me in the most challenging moments of my research and writing, and her enthusiasm, kindness, and optimism have been an enormous boost for me when I needed them most. She inspired me to pursue new heights.

I am very grateful to the faculty and staff at the School of International Relations and the Department of Political Science – Linda Cole, Cathy Ballard, Jody Battles, Karen Tang, Veri Chavarin, Wayne Glass, Carol Wise, John Odell, Laurie Brand, and all the faculty whose courses I took. The School of International

Relations, the Center of International Studies, and the Politics and International Relations program have generously supported my research and studies, enabling me to present my research at major international conferences and to attend research and training courses such as the Azerbaijan Diplomatic Academy Energy School, the University of Essex Summer School (UK), and the Consortium on Qualitative Research Methods (CQRM). I am also thankful to USC Center for International Business Education and Research (CIBER) for research awards and the Unruh Institute of Politics for the grant allowing me to intern at Global Green USC in Washington, DC.

I thank my family and friends in Russia – Mikhail Mumlev, Julia Kucherova, Olga Ovcharuk, Lena Studenkova, Tatyana Gubaidulina, and Olga Rybova. Your support has been my foundation and my strength.

I devote this book to my parents Vladimir and Lubov Svyatets, and my sisters, Nadia and Tanya, who have always encouraged me to be brave and to explore the world.

Introduction

Why are bilateral relations, especially in the area of energy security, so different in the cases of U.S.–Russia, U.S.–Azerbaijan, and Russia–Germany energy deals? Why do some states find common ground despite differences, while others, with all the seemingly favorable conditions, are sinking into animosity? This study explores the varying outcomes of energy cooperation, defined as diplomatic relations, bilateral trade, and investment in oil and natural gas. The international relations literature – broadly speaking, realism, liberalism, and domestic politics – has long attempted to come up with a one-line explanation of energy security and cooperation. It has revealed narrow and one-sided aspects, not embracing the full complexity of energy issues. Using the existing explanations as a starting point, this study offers a systematic approach that incorporates three main aspects of energy security decisions: economic potential, geopolitical rivalry, and the interests of domestic groups.

This study concludes that if the economic potential (defined by geographic proximity and resource availability) is very high, such as in the case of Russia–Germany, states can overcome geopolitical rivalries and historical enmities in favor of energy cooperation. However, if the economic potential is relatively low (because of geographic obstacles or easily available alternative suppliers, as in the cases of U.S.–Russia and U.S.–Azerbaijan), then geopolitics prevails – for example, decisions to bypass Russia when building pipelines or to limit American access to contracts in Russia when U.S.-Russian relations are strained. In all the cases explored here, domestic interest groups have mixed influence: if they are united along energy issues, they usually successfully achieve their energy policy goals, although the impact of these groups often becomes intertwined with state interests. In other situations, when powerful interest groups are divided or focused on non-energy-related issues (such as ethnic priorities), their influence over energy deals is much lower.

Using primary sources, interviews, and field research in Russia and Azerbaijan, this study empirically explores the cases of U.S.–Russia, U.S.–Azerbaijan, and Russia–Germany energy cooperation. The topic of energy security is highly dependent on current events and the changing context of the international system. This book is no exception. In addition to events of the Cold War and post-Soviet conflicts such as the Russia–Georgia war of 2008, one chapter is devoted to the Russia–Ukraine conflict 2014–2015, which has been a pivotal event that has affected Russia's relations with the West. This study explores the effects of the Ukrainian February revolution of 2014, which preceded the military conflict in Eastern Ukraine, on the energy security relations between Russia and the West.

Most of the energy deals between the West and post-Soviet countries have happened in the area of oil and natural gas. Both are non-renewable resources, strategically vital for every country's survival. Production of oil and natural gas is normally very oligopolistic in its nature. Crude oil has a high Energy Profit Ratio (the amount of energy produced from a certain amount of fuel) and is more prevalent for transportation use than other types of fuel, including hydrogen, ethanol, condensed natural gas. Oil is also relatively easy to transport by pipelines and tankers. That is why oil will most probably stay a major fuel source for decades, only slowly being replaced by natural gas and renewable sources. Natural gas, in its turn, is a transition fuel that is prevalently used for electricity production. It is cleaner than oil, as natural gas emits less carbon dioxide than oil when being burnt. However, it is not as transportable between continents, because its liquefaction requires new extensive infrastructure and additional investments and costs. That is why natural gas has been mostly produced and transported by pipelines locally and regionally (for example, from Russia to Western Europe).

Despite the great progress in seismic exploration methods and drilling equipment, consensus is still lacking among oil and gas experts about how much oil is available for production around the globe and in each specific country and how long the supply will last. Competing models disagree even on the 'limitedness' of oil, from the Hubbert model that claims that resources are about to run out to the Cornucopian model that claims that the resources are unlimited (Blanchard, 2005, pp. 8–21). However, it is agreed that oil and natural gas are non-renewable (i.e. finite) resources. Shale oil and natural gas exploration by hydraulic fracturing (fracking) has postponed Hubbert's peak in many countries, and especially in the United States, making the United States one of the largest producers of oil and natural gas again. This new technology, which made plentiful oil available and affordable, has been controversial in its environmental effects (the impact on aquifers, earthquakes, methane leaks, waste water spills, and others), but it has certainly ensured a stable supply of oil and natural gas for the American market.

Reliable, affordable, and sustainable supply and demand of energy sources are what lies in the core of the definition of energy security (Bressand, 2009; Shaffer, 2009). Simultaneously, energy security does not automatically imply *energy independence* from other countries, i.e. a nation possesses sufficient energy resources without the need to import them. Energy independence is not always possible because of geographical determinants, but energy security can be achieved through mutual investment, technology transfer, inter-governmental support of energy deals, and trade in energy resources.

Trade and investments usually mean mutual dependency between energy consumers and energy producers. Traditional energy security literature mostly focuses on the vulnerability of energy consumers who lack their own resources, such as Ukraine in this study. However, energy producers face their own challenges, such as environmental problems, volatile oil prices, and the dependency on consumers and transit countries, as well as the need to build expensive infrastructure (pipelines, storage facilities, and other elements). The extant literature categorizes all states as either consumers or producers, which is an oversimplified theorizing, because most countries are, in fact, producers and consumers of energy resources simultaneously (Shaffer, 2009). This study addresses this mutual dependency and looks more closely into the energy mix of the involved states.

Each nation, when lacking energy sources domestically, imports them from global and regional markets. Especially for oil, energy markets are constantly shaped by the forces of globalization and regional segmentation, and major international energy companies have to follow those developments. The companies are becoming more integrated, combining oil and natural gas extraction ('upstream' in the oil and gas industry terminology); processing, refining, and retail ('downstream'), and transportation ('midstream') ('OGA North America Oil and Gas Industry Research Guide', 2014). Previous divisions between oil production and natural gas production have also become blurred, because usually they go hand in hand.

In addition to being more vertically integrated, oil and natural gas production has become more regionally oriented (e.g., North America, Eurasia, and the Middle East) in order to minimize the transportation costs (Shafranik, 2009, p. 26). It is especially obvious in the natural gas trade. The natural gas trade has been revolutionized with the invention of liquefied natural gas (LNG) technologies that are making the natural gas trade more globalized. LNG is now moved around the world by tankers, making natural gas a truly global commodity. Technology is steadily globalizing what traditionally has been a regional trade issue. However, the new LNG technologies are still quite expensive, and transportation costs are high. That is why natural gas (in its usual gas form) is still delivered via pipelines between neighboring countries, under set conditions, negotiated by the governments, including a stable price for the period of a contract.

Prices of oil, however, are not stable, and they affect both producers and consumers when they fluctuate. Oil prices are very volatile. The biggest oil producers affect global oil prices by reducing or increasing their oil production and/or through direct agreements within such organizations as the Organization of the Petroleum Exporting Countries (OPEC). Even non-OPEC countries sometimes coordinate their actions with OPEC: for example, in 2009, the Russian Deputy Prime Minister, Igor Sechin (responsible for energy and environment) invited OPEC representatives to Moscow to discuss oil prices ('Sechin: Rossiya Origlasit Predstavitelei OPEC v Moskvu', 2009).

Oil states adjust their oil and gas production volume in order to keep prices at a certain level. In the global oil market, Saudi Arabia has long played the role of a swing producer, which is able to quickly increase or decrease production because of the country's enormous proven reserves of oil (Yergin, 1992). Other countries, such as Russia and the United States, possess a more limited leverage on global oil prices because they produce oil almost at their capacity ('Rossiya Umen'shaet Svoi Export Nefti', 2009).

In addition to oil production volume, oil prices are affected by the changes in information about existing oil reserves and new discoveries, showing if the market believes that oil production will substantially increase in new oil fields. Leading experts in oil and gas research warn us, however, that information about global oil and gas reserves might not be accurate, even at the current level of exploratory technology. Roger Blanchard, the author of *The Future of Global Oil Production: Facts, Figures, Trends and Projections*, argues that data on oil reserves may be misleading and overly optimistic, partly because governments are motivated to inflate their numbers. Governments use different methods for calculating oil and gas reserves. There is also an interchangeable use of the terms *proven* reserves and *probable* reserves (the latter are usually significantly higher than the former). Finally, crude oil production is often accompanied by other *liquid hydrocarbons* (condensate, natural gas liquids, refinery gain, and other hydrocarbon liquids) (Blanchard, 2005, pp. 1–4). When the by-product condensate and other substances are included in the data on crude oil reserves, the figures look much higher.

Adding to conventional reserves, the global energy mix has been recently redefined by the 'shale oil and gas revolution' in the United States and other countries. Since 2013, this 'revolution' has propelled the United States to the top three oil and gas producers in the world. Hydraulic fracturing, used for extracting shale oil and gas, has eliminated the United States' need for imported natural gas, and somewhat decreased the volume of imported oil, making the country self-sufficient in natural gas and less vulnerable in oil. Moreover, the United States has become an exporter of natural gas, which was unthinkable just 10 years ago. Those changes in the American energy mix substantially affected the dynamics of the relations with other energy producers, such as Russia and Azerbaijan.

All these oil- and gas-related factors, combined with geopolitical aspects and domestic groups' influence, makes studying energy cooperation very interesting, especially in such globally-relevant cases as the U.S.–Russia, U.S.–Azerbaijan, and Russia–Germany. In the following chapters, the unified framework of geopolitics, economics, and domestic lobbies is applied to the cases.

Chapter 2, 'U.S.–Russia energy ties', looks at the development of the Russian energy sector and foreign investments since the 1990s. It reveals a gradual decline in cooperation, accompanied by growing political hostilities. Based on the availability of oil and natural gas, one might expect a higher level of U.S.–Russia energy trade and investment, considering that the United States is the biggest consumer and Russia is the biggest producer of oil in the world. Chapter 2 explains the lack of cooperation between the two countries, starting with the economic dimension.

The economic climate analysis is followed by the close examination of U.S.–Russia geopolitical factors, such as the legacy of the Cold War thinking by policymakers, contemporary military rivalries in Eurasia, security relations with other countries, and the aspiration to access the world's energy resources. These geopolitical factors proved to be extremely powerful, directly affecting the outcome, such as animosity and the low level of bilateral trade and investment. Those geopolitical factors have been affected and exploited by domestic interest groups in Russia: the Russian political elite, the Russian energy lobby, and American lobbies in Russia.

As a specific example in how these factors interplay, the Yukos case is examined in Chapter 2, showing how economic, geopolitical, and domestic factors led to a crash of the biggest oil company in Russia and the political annihilation of its owner Mikhail Khodorkovsky, who boldly attempted to merge his company with one of the U.S. majors. As a result, Yukos was sold in pieces to Russian state-controlled energy companies, and Khodorkovsky himself ended up in jail for a decade.

Chapter 3 explores an almost opposite case of growing energy security ties between the United States and Azerbaijan. The United States government and American companies have been actively involved in Azerbaijan's economy and energy trade, despite Azerbaijan's distant, landlocked geographical position.

The analysis starts with Azerbaijan's economic and political transformation in the 1990s and an influx of foreign investments to the country. During that decade, the economic and political climate in Azerbaijan and Russia were similar. Major divergences started during the 2000s, when, unlike Russia, Azerbaijan created a very favorable investment climate for foreign investors, especially in the energy sector, working productively with very diverse countries (Russia, the United States, Iran, Turkey, etc.). As a result of growing investments, new oil and gas reserves were discovered and the recovery rate was significantly improved, helping the production level soar.

Azerbaijan has navigated successfully among the geopolitical rivalries in the Caspian, where great powers have competed for influence and access to energy resources. The contracts between Western governments and Azerbaijan, such as the Baku–Tbilisi–Ceyhan pipeline project, have been implemented despite some resistance from domestic lobbies representing other countries, such as Armenia. The Armenian lobby in the United States has not been able to prevent the BTC pipeline implementation or to divert it through Armenia, even though the lobby has been powerful on other issues. The U.S. government has valued Azerbaijan as one of the key partners in the Caucasus in the U.S. policy of multiple pipelines, which aims at reducing the Russian domination of European energy markets.

Chapter 4 addresses the third case study: Russia–Germany energy cooperation. It is an interesting case, in which the outcome is somewhat unexpected. Despite the legacy of the Second World War and the Cold War, contemporary Russia and Germany have managed to achieve an unprecedented level of cooperation, in which oil and gas trade and investment have been the major focus. It is true that the two countries' cooperation is favorably conditioned by their geographic proximity, but geography alone cannot account for why the two states have been so intertwined. Other neighboring states, such as Poland and Estonia, have not cooperated with Russia as much as Germany, despite their own geographical position.

Chapter 4 explains how the quintessential EU member – Germany – decided to defy the EU energy policy of reducing dependency on Russia. Instead, Germany and Russia built a pipeline that directly connected the two countries. The pipeline on the bottom of the Baltic Sea – Nord Stream – has been estimated by the German and Russian governments as economically feasible, and Vladimir Putin's personal ties to Germany have also played a significant role. In addition to providing Germany with unprecedented access to Russian natural gas supply, the Nord Stream pipeline has substantially reduced the geopolitical influence of transit states, such as Poland and Ukraine.

It is not to say that Ukraine has lost its importance completely, and Chapter 5 looks at the recent domestic events in Ukraine and the recent conflict between Ukraine and Russia. Even though Russia and Ukraine have had disagreements over natural gas prices several times over the last 10 years, the conflict in 2014–2015 has become the lowest point in the history of the two states' bilateral relations since the breakup of the Soviet Union. The enmity in the energy domain was accompanied by other major events that shook up the post-Cold War international system. After the Ukrainian February revolution of 2014, Russia helped the region of Crimea hold a referendum, in which the voters supported Crimea's leaving Ukraine and joining Russia. The referendum has been strongly opposed and contested by the West.

After the referendum, Ukraine's Eastern part (the Donetsk and Luhansk regions) started its own secession movement. The Ukrainian government has used military force to suppress the self-proclaimed republics and to protect Ukraine's territorial integrity. The Western governments and media presented evidence that Russia sent military assistance to the separatists in Donetsk and Luhansk, claiming that Russian soldiers without insignia were fighting there. Those accusations were repeatedly denied by the Russian government. Amidst the hostile rhetoric on both Ukrainian and Russian sides, connecting the conflict to the legacy of WWII and revisiting historic events, this tragic conflict has led to more than 5,000 casualties and hundreds of thousands of refugees who fled to Russia. Not surprisingly, these events have direct implications for European energy security.

The final chapter contains conclusions and policy recommendations, focusing on the ways to reduce the dependence on hydrocarbon fuels that have long fueled international conflict and environmental decline.

References

Blanchard, R. D. (2005). *The Future of Global Oil Production: Facts, Figures, Trends and Projections, by Region*. Jefferson, N.C: McFarland & Company, Inc.

Bressand, A. (2009, July 6). Global Trends in Energy Production and Consumption: An Outlook for Energy Security and Governance. Lecture at the Energy Politics and Economics Summer School.

OGA North America Oil and Gas Industry Research Guide (Q1 2014) – Analysis of Upstream, Midstream and Downstream Infrastructure, Investments, Companies and Outlook to 2025. (2014, September 4). *PR Newswire*.

Rossiya Umen'shaet Svoi Export Nefti. (2009, March 12). Retrieved March 21, 2009, from http://www.gazeta.ru/news/lenta/2009/03/12/n_1340154.shtml

Sechin: Rossiya Origlasit Predstavitelei OPEC v Moskvu. (2009, March 12). Retrieved March 21, 2009, from http://www.gazeta.ru/news/lenta/2009/03/12/n_1340147.shtml

Shaffer, B. (2009, July 5). Lecture to the Energy Politics and Economics Summer School Participants.

Shafranik, Y. (2009). Globalnaya Energetika i Rossiya. [The Global Energy Sector and Russia]. *Vestnik Aktualnikh Prognozov. Rossiya: Tretye Tisyacheletije*, *21*, 25–27.

Yergin, D. (1992). The Prize: The Epic Quest for Oil, Money, and Power. New York, NY: Simon & Schuster.

1 Energy security

The current approaches of scholarship

Energy: trade or war?

Energy is one of the most central issues in contemporary global politics. It is frequently at the heart of global trade issues, because energy prices are often the key factor in the differences between growth and prosperity or shortage and recession. Energy is vital for developed countries to maintain growth and living standards, and no less central to the world's developing economies. From the Organization of the Petroleum Exporting Countries' (OPEC) embargo of the 1970s, which sprang from the conflicts in the Middle East and sent Western economies into prolonged decline, to numerous confrontations of the last two decades in the Caucasus, Central Asia, Africa, and elsewhere, energy security has had a great impact on great power relations. There is scarcely a major international issue not directly or indirectly linked to energy: whether it is the 'resource curse' distorting the domestic politics of individual states; resource competition provoking civil or regional conflicts; the maneuvering of major powers to secure energy independence or geopolitical dominance; or the climactic dangers of fossil-fuel use. This study of energy security and cooperation obviously cannot encompass all of these issues. Its focus is narrower, on the questions of energy relations between the West and the post-Soviet regions – the former a major consumer and importer, as well as a producer, and the latter a major producer and exporter.

Major consumers of oil (the United States, Europe) do not produce enough to ensure their self-sufficiency, and thus they have to secure supplies from other regions of the world. One important oil- and gas-producing region is the Former Soviet Union (FSU): Russia alone possesses about 60 billion barrels out of the world's proven reserves of 1,200 billion. Kazakhstan and Azerbaijan follow with 30 billion and 7 billion barrels (EIA, 2012). The numbers are even more impressive for natural gas: the FSU, primarily Russia, possesses 40 per cent of the world's natural gas reserves. However, access to these resources has not been easy, and many FSU countries remain a risky area to do business in because of tight government control of resources and undeveloped domestic political and economic institutions.

Post-Soviet states differ in their relations with the West. For example, Azerbaijan, despite somewhat limited oil and gas reserves, has been active in its cooperation with the Western governments and major oil companies. Russia, on the contrary, has had complicated and limited relations with the United States, particularly in the last decade. However, Russia has closely cooperated with Germany, despite the European Union's attempts to diversify its energy suppliers and reduce Russia's influence.

These various paths of relationships have led to the following research questions: *Why does energy cooperation (trade and investment) succeed in some cases but fail in others? What are the determinants of energy cooperation and energy security?*

Similar questions have been investigated by the literature on energy security. This aspect of international relations is a multidisciplinary area that borrows elements from political science, economics, environmental studies, as well as geology and natural sciences (Aalto, 2008b, p. 5; Wu, Fesharaki, Hosoe, & Isaak, 2008). Each body of literature provides an insight into important aspects of energy security. However, they tend to do so in isolation, to the exclusion of factors central to other approaches. In other words, the literature offers few if any integrated approaches that simultaneously take into account many of the important factors and systematically explain different outcomes of energy relations. This study offers such an integrated approach to energy security.

To explain why some energy projects, seemingly less economically justifiable, become implemented, while others are neglected, three major approaches in the international relations literature – realism, liberalism, and domestic politics – are applied here for the proposed unified framework. Realism argues that trade in strategic resources is strictly determined by national interest, in which a state's overriding concern for survival and power in the international system is highly prioritized. Liberalism, by contrast, stresses the benefits of cooperation, particularly economic, and therefore the promise of easing confrontation for mutual gain. Both liberalism and interdependence (expressed in trade, investment, mutual benefits, norms, and institutions) can be described as an optimistic view of international relations, while realism or statism (focused on the balance of power, resource competition, energy nationalism, and vulnerability) offers a more pessimistic view (Tanner, 2009). Finally, the domestic politics approach highlights the influence of various lobbies and interest groups that have a stake in foreign policy issues, such as energy projects. These groups, if powerful and organized enough, can shape states' policies in directions that contradict the 'national interest' or 'objective economic benefit' to the gain of their influential constituencies.

The best way to test those theories and the unified theoretical framework is to apply them to carefully selected case studies. The bilateral cases of U.S.–Russia, U.S.–Azerbaijan, and Russia–Germany have been chosen here for several reasons. First, these cases ensure variation in the outcome of energy cooperation (the 'dependent variable') and incoming factors, such as economic climate, geopolitical rivalries, and domestic lobbies ('independent variables'). Second, each of them presents a challenging case of energy cooperation, based on resource availability and geographic determinants, domestic political institutions, or historical enmities. Finally, the cases include prominent consumers and producers of oil and gas, thereby ensuring policy relevance.

International relations scholarship, applied to energy security: history of thought

Inspired by the extant literature of international relations, this study offers a unified explanation applicable to energy security and cooperation. Broadly defined, the framework here looks at geopolitical differences and calculations (national interests of states), economic motives, and the lobbying of domestic groups. For scholars and practitioners, it might be beneficial to learn about the roots of energy security theorizing in the international relations literature, as those theoretical works have often had a direct implication for governments' policies and decisions. These approaches are then integrated in this study, systematically analyzing the cases of U.S.–Russia, U.S.–Azerbaijan, and Germany–Russia economic and political ties in the area of energy security.

Is it all about military power and national interest?

The first traditional approach – realism – argues that a state is a unified entity and that domestic interest groups comply with the goals of the state. Cooperation between states, including trade in strategic resources, occurs only if it advances states' national interests, in which security considerations prevail over profit maximization.

A prominent representative of this approach – Stephen D. Krasner – in his *Defending the National Interest* examines how the state, as a unified and autonomous actor, builds relationships with multinational private corporations in formulating resource policies. Krasner draws on the classic realist work of Hans Morgenthau and his view of states' national interest overwhelmingly defined in terms of power (Morgenthau, 1948). The objectives of the state, Krasner continues, constitute the national interest, which is a consistent set of goals defined by policy-makers. Comparing the statist approach to interest-group liberalism and Marxism, Krasner argues that statism provides a better understanding of raw-material-relations than the other two. He defines the state as a 'set of roles and institutions having peculiar drives, compulsions, and aims of their own that are separate and distinct from the interests of any particular societal group' (Krasner, 1978, p. 10). In his view, the state is a rational homogenous decision-maker, in raw material investments and in all other matters, and the government defines the national interest. While the state's insulation from societal pressures is disputed by other theoretical approaches, Krasner's theoretical foundations of statism have had a profound influence on other resource-policy studies.

Another influential scholar of this theoretical school – Robert Gilpin – continues this line of argument and argues that 'in a highly integrated global economy, states continue to use their power and to implement policies to channel economic forces in ways favorable to their own national interest and the interests of their citizenry' (2001). Gilpin adds that free-market mechanisms are not sufficient to ensure the world economic system's smooth operation. It is especially important for the global oil market, in which oil-possessing states control oil production volume and the access to oil fields on their territory. Gilpin's definition of political economy – a 'sociopolitical system composed of powerful economic actors or institutions such as giant firms, powerful labor unions, and large agribusinesses that are competing with one another to formulate goverment policies' – is highly relevant for the energy sector (2001, p. 38). Gilpin is responding to liberal interdependence theorists, such as Robert Keohane and Joseph Nye (1989), as well as to the literature that emphasizes the growing power of global corporations. Gilpin's argument is that states are still the ultimate actors in world politics, and they shape even the dealings of powerful multinationals; if states do not want a deal implemented, they will halt it.

Other authors have focused specifically on oil-rich states, in which oil has played a role beyond that of a regular commodity. In those countries, the oil industry has been more politically and economically influential than most other industries. One of the best historic accounts of the oil industry's development globally and its use by governments as a tool of power is Daniel Yergin's *Prize* (1992), in which the author develops three main themes. The first theme is the rise of the oil industry as global business, which has the biggest impact on modern societies. The second theme focuses on how this industry is 'intertwined with national strategies and global politics and power'. The third theme concentrates on how mankind has developed into what some experts call the 'Hydrocarbon Man' (Yergin, 1992). Yergin's volume provides knowledge and insight into how governments promote their national interests though resource diplomacy and war.

Yergin's latest book, *The Quest* (2011), examines more recent developments in energy power politics. One of them has been an increasing globalization of the world economy, including the energy industry. One of Yergin's main cases is Russia, which has become a major oil and gas producer, rivaling Saudi Arabia. Prior to 1991, oil revenues helped finance Russia's imports of consumer goods and the ailing economy; the oil industry was the major economic force throughout the Soviet period. Sinking oil prices in the 1980s, however, contributed to the economic collapse of the Soviet Union. Yergin provides an overview of the oil industry transformation in Russia from the Soviet model to market capitalism that has now turned again toward increased state control (Yergin, 2011, pp. 31–37). During his first and second term, in 2000–2008, President of Russia Vladimir Putin announced that oil and gas were the key to Russia's economic power and its place in the world economy. Since then, energy resources (natural gas, nuclear energy, oil) have been a major 'tool' of Russia's foreign policy, especially with the neighboring states (the former Soviet Union republics, Iran, China, and even the European Union). Contemporary Russia is an example of how government is intertwined with the energy industry. Both in the Soviet and post-Soviet eras, Yergin sees Russia as a quintessentially realist state in its energy politics.

Russia is not the only country where the energy sector is largely in the hands of the state: out of the world's twenty largest energy companies, sixteen are state-owned. State ownership is considered by governments of oil-possessing countries – including some largely capitalist free-market countries – as beneficial for controlling their reserves and revenues from oil and gas export. Russia, after the sweeping wave of privatization of the 1990s, quickly returned to the state-controlled model, increasing the share of government-controlled companies from 13 per cent in 2004 to 40 per cent in 2007 (Pleines, 2009, p. 71). The last decade was the period when Russia started to move away from the Western political influence, becoming more resurgent.

The realist-statist approach views energy security as a zero-sum game and a constant rivalry. This approach relies heavily on the notion of geopolitics as 'a subfield of political science and geography… [that] focuses on the relationship between territory and power, particularly the influence of geography on state behavior' (Dekmejian & Simonian, 2003, p. 5). Energy resources are often discussed as a 'weapon' for geopolitical domination and manipulation. For example, analyst Vladimir Socor refers to Russia's pipeline projects in Eurasia as a 'blitzkrieg capture', with the goal to dominate the European oil and gas market (Socor, 2008). A similar theme is presented in Marshall Goldman's *Petrostate: Putin, Power, and the New Russia* (2008) – a very detailed account of Russia's energy policies since the mid-nineteenth century, with a special emphasis on the last decade. In his analysis, Goldman goes beyond geography and shows how oil and gas are used as a political and economic weapon, potentially able to destabilize transit countries in the Caucasus region and Central Asia (Goldman, 2008, p. 149).

Many other studies have also concluded that energy resources can be a reason for conflict or outright war. Mary Kaldor, Terry Lynn Karl, and Yahia Said, in their book *Oil Wars* assert that oil causes, exacerbates, or mitigates wars and affects the nature of a conflict (Kaldor, Karl, & Said, 2007). The authors analyze the cases of Angola, Azerbaijan, Colombia, Indonesia, Nigeria, and the Chechnya war in Russia, concluding that oil money finances these conflicts, in which domestic groups compete for oil resources in these geopolitical conflicts related to oil. Such wars can potentially happen, as some authors (Standlea, 2006) argue, even in the Arctic: the area that contains enormous energy resources and becomes more accessible as the global warming progresses. Notwithstanding their many variations in subject or emphasis, these studies share realism's consistent skepticism about the liberal assumption that trade in energy resources among states would bring more peace and stability.

Other authors explore the effects of terrorism on the oil community, including special terrorist tactics, attack methods, shocks resulting from oil supply destruction, global choke points in oil delivery, and other possible damages to the industry (Adams, 2003). An in-depth analysis of potential weapons and tactics that can be used by terrorists (bombs, explosives, kidnapping, hijacking, dirty bombs, chemical weapons, and others) provides a valuable overview of the security dimension of various energy projects, such as pipelines, tankers, and rail delivery. In the last decade, physical security of energy projects has become an important part of their planning and execution.

In sum, the realist–statist approach maintains that security considerations, the position of a state in the international system and the state's influence in a region, consistently prevail in foreign, including energy, policy making. Energy cooperation is likely if it is in the national interest – defined in terms of power. When the national interest is at stake, economic profitability is usually sacrificed. Energy policies are considered by this approach as a tool of domination and conflict, and thus compromise is difficult to achieve.

Economic liberalism and interdependence

Unlike realism, liberalism argues that even in the area of strategic resources, states should be able to trade with each other for mutual benefit. The literature on economic liberalism and interdependence focuses on the *economic* fundamentals of trade and cooperation, which often lead to interdependence between countries.

Interdependence between two or more states is believed to contribute to peaceful relations among them. A pair of states does not have to 'like' each other in order to trade: governments are supposed to be able to put their differences aside for the sake of their access to valuable resources, markets, and mutual profits (Axelrod & Keohane, 1985). States prefer to trade instead of invade each other (Keohane & Nye, 1989). Realists would probably prefer to go to war to escape dependency, while liberals argue that mutual dependency reduces the chances of war. Unlike realism, that emphasizes 'dangerous dependencies' and 'energy supremacies' of oil and gas producers, liberalism points to interdependencies that foster cooperation (Copeland, 1996). Political economists study how free trade influences domestic economies and political systems; a common prescription is that resource-rich countries have a greater chance to become wealthy if they commit to free trade (Sachs & Warner, 1995).

In contrast to realists who worry, for example, about Europe's dependence on Russian oil and gas, liberal interdependence theorists look at Russia's dependence on Europe's consumption of Russian resources. In this case, they would view Russia's dependence on Europe for oil and gas consumption as a guarantor of long-term trade, investment, and peace. Economic interdependence theory predicts that pipelines can be a 'peace' catalyst, even during such difficult gridlocks as the Nagorno–Karabakh dispute between Azerbaijan and Armenia and the Georgia–Russia war over South Ossetia (Closson, 2009).

Energy economics scholars explain the trade mechanisms, often neglected by realist scholars. They emphasize the role of geography in assessing energy projects and transport routes. Landlocked countries, for example, have limited access to foreign markets. The energy economics literature also explores the efficient ways of exploration and trade of global energy resources; the quantity and quality of a country's oil, gas, uranium, and other resources, as well as the rate of their depletion, all of which define the feasibility of a project (trade, investment) and of resource distribution (Falola & Genova, 2005). Such research provides us with valuable insight into energy project financing, trends, and tendencies in the world's oil production, economics of various sources of energy (oil, gas, coal, nuclear, renewables) and money markets related to energy – again, the areas usually not studied enough by realism (Banks, 2007).

Energy economics research is more focused on the laws of trade and investment (often expressed in mathematical formulas), rather than politics, although these scholars also examine several regional energy organizations (such as OPEC) and use case studies, as well as the economic history of resource use. They often conclude that economic developments in the energy industry shape global energy markets that are dominated by private and state energy companies and institutions (Falola & Genova, 2005).

Economists also study historical patterns and distortions that often persist in oil and gas trade. For example, former World Bank chief economist Joseph Stiglitz, in his *Globalization and Its Discontents*, touches upon economic and price distortions in the post-Soviet countries. During the Soviet period, prices were dictated by the command economy, not by market mechanisms. After the dissolution of the USSR, Russia continued to provide oil and gas at below-market prices to former Soviet countries in order to 'purchase' their loyalty and support. Russia could afford these price distortions only thanks to the country's huge energy resources (Stiglitz, 2002, p. 139). Timothy Colton also explores this issue, noting that the Yeltsin administration provided oil and gas to post-Soviet states at Soviet-era discounted prices in order to keep their economies afloat and to prevent economic collapse and social unrest in those countries (Colton, 2008, p. 266). Such research has been especially relevant during the recent events in Ukraine in 2014, when Russia–Ukraine negotiations were focused on the same issues that Stiglitz was writing about 13 years ago.

Energy economics provides very detailed accounts of oil and gas reserves and energy companies' activities. One such study, Jonathan Stern's *The Future of Russian Gas and Gazprom*, examines important details about Russia's natural gas resources and how they have been explored and developed over the last two decades, including major players in the industry, geography, and a wealth of statistical knowledge (Stern, 2005).

In sum, liberalism stresses the importance of profitability, free markets, and non-state actors, searching for common interests in order to reach mutually profitable solutions. In the classic contrast with realism and its focus on *relative* gains, the liberal international relations approach emphasizes *absolute* gains: if everyone profits, it is not so crucial who might profit more and who less (Grieco, Powell, & Snidal, 1993). Interdependence between states leads to better relations and peace, as both states find mutually profitable solutions for their needs.

Domestic interest groups: what is their role?

The relevant literature on domestic politics and interest groups, pioneered by the constitutional-legal and the group approaches (Baumgartner & Leech, 1998), is centered on American politics and influencing the U.S. Congress. This literature provides insight on how interest groups shape foreign (and energy) policies.

The origins of the domestic interest groups literature go back to the 1950s, when pluralism theories were developed by David Truman, Robert Dahl, and Jack Walker, among others. A founder of the pluralist school, David Truman, explores interest groups' role in the political process, based on their conflicting and competing interests. Truman lists several factors that determine the power of interest groups. First, their 'strategic societal stance' is determined by their status and prestige, closeness to government structures, and their access to valuable information. Second, their success is shaped by such internal features as the degree of formalization, leadership abilities, and financial resources. Finally, interest groups' ability to exist and function is dependent on the democratic structure of the government, with its civil liberties and human rights (Truman, 1951).

Another branch of this literature argues that it is not interest groups but political parties that are more important and efficient for citizens' promoting their interests in democracies, which are a 'competitive political system in which competing leaders and organizations define the alternatives of public policy in such a way that the public can participate in the decision-making process' (Schattschneider, 1960, p. 141). When political parties are strong and organized, they are able to project the population's ideas to the government. Diverging from the pluralist school, these scholars conclude that political parties are more than mere aggregates of interest groups.

Scholars have long explored the best ways to encourage people to work together and pursue their common interests. Such research became very prominent with Mancur Olson's *The Logic of Collective Action*, which explores what makes group members cooperate for their collective interests. The premise of his theory is that rational and self-interested individuals join, via either incentives or coercion, groups of common interests. Group size matters, too, and Olson finds that it is more difficult to motivate members in a larger group (Olson, 1965). Unlike the previous group theory by Leon Festinger and Harold Laski, which assumes that groups of all sizes equally attract members whose participation is voluntary and universal, Olson concludes that it is much harder to provide public goods and to achieve collective bargaining in a larger group because of the increasing sub-optimality. Thus, larger groups are less effective to further their members' common interests.

By the early 1970s, after Olson's breakthrough volume, the literature on domestic politics continued to focus largely on collective-action dilemmas, alongside such topics as lobbying methods, interest-group democracy, political action committees, and electoral politics. Multiple scholars of international political economy have studied how the electorate influences trade policies (Foyle & Belle, 2010): among them, Grossman and Helpman (2002) and Bueno di Mesquita (1997). These authors look at the structure of the electorate and the divisions in governments to predict trade policy openness and protectionism. Constituents lobby the government through public interest groups, labor unions, and other grassroots movements, funded by donations and grants (Wolpe & Levine, 1996, p. 2).

Lobbying is one of the main methods used by domestic groups to shape policies, such as, for example, of the U.S. Congress (Carter and Scott, 2004; Rubenzer, 2008). The definition of *lobbying* has varied in the literature, and scholars argue what needs to be included in this concept. A narrow definition only includes professional lobbying activities: supporters of the narrow approach (e.g., Russian political scientists Avtonomov, Neschadin, Blokhin, Vereschagin, Grigoriev, and Ionin) assert that only non-governmental groups and businesses can be categorized as lobbyists (Tolstykh, 2006, p. 9). The other school of thought pushes a broader definition: lobbying can be done by any societal entity (including government agencies) that tries to promote a particular issue and a policy (Tolstykh, 2006, p. 8).

In the United States, lobbying has been defined and systematized in several categories. First, there are *contract* lobbyists, hired to promote somebody else's interest for a fee or salary. In Washington, DC, a lot of such firms are located on the famous K Street. They usually specialize in a certain issue-area, based on their expertise and networking. This category is different from *corporate lobbyists*, who are companies' employees that represent corporate interests. Their voices are very influential, as corporations provide jobs and tax revenues in electoral districts. Yet again, corporations or industries join *business and professional associations* that 'represent the collective interests of an industry or group of industries' and are de-facto lobbying organizations themselves (Wolpe & Levine, 1996).

Professional lobbying companies in the United States, by law, are required to be transparent about their donations and activities. Their names and phone numbers are available in special reference books. Their financial contributions to political campaigns and other activities are supposed to be publicly known, especially via new media technology and online access to government legislative documents.

This transparency is not always the case in other countries. For example, in Russia and Azerbaijan, lobbying activities happen behind closed doors, and there are few ways to limit financial contributions to politicians or even to find out about those donations. Moreover, lobbying as such has a negative connotation in the post-Soviet regions, as lobbying was almost 'an insulting word' in the Soviet era (Tolstykh, 2006, p. 28).

Each domestic group, when promoting their interests, uses their 'narratives', which means political and economic 'story-telling'. They are especially common 'when scientifically and technically secure information is not available, as it most often is not, and when policies nevertheless need to be developed, decisions made and funds assigned' (Aalto, 2008a, p. 29). Such 'narratives' are used alongside (and intertwined with) political analysis and prescriptions produced by think tanks and research institutions. In the United States, such think tanks as the Brookings Institution, the Carnegie Endowment for International Peace, the

Heritage Foundation, and others are influential in the government's foreign policy decisions.

Ethnic lobbies also promote their interests via narratives in their lobbying mechanisms, using historical parallels and their economic interests (Foyle & Belle, 2010). Of particular relevance to this study are the Armenian, Polish, and other ethnic lobbies in the U.S. and Western Europe whose emphasis on historic victimization is coupled with the current policy positions on issues from NATO expansion to pipeline construction.

Similar actors, interests, and narratives are found in the energy sector, as well. Energy projects and deals can only be implemented if domestic actors in participating countries support them (or, at a minimum, are unable to block them). Domestic actors include energy 'movers' (producers and transportation companies), energy industry shapers (the 'financial, research, social, and supporting organizations that facilitate the movement of energy from producers to users'), energy users (consumers), and energy regulators (government agencies) (Geri & McNabb, 2011, p. xxix).

Domestic groups that followed the energy policy line of their respective governments usually achieve their goals more easily. The overlap between the interests of domestic groups (voluntary organizations, membership groups, associations, institutions, social movements, special interest organizations, pressure groups, lobbyists, political action committees (PACs), and many others) and government agencies results in the groups' conformity to the government agenda (Lowery & Brasher, 2004, p. 5). These groups are very visible and vocal in developed democracies, but they also have their influence (not as visible) in non-democratic states. In non-democratic states, interest groups may have an even more direct influence on state policy-making.

The literature has followed these differences between democracies and non-democracies, as well as the prominence of domestic groups in the economy. Such prominence in policy-making and special studies, applied to the energy security domain, has followed the pattern of energy crises: in the 1980s, during the oil crisis, the groups became vital and salient. However, in the 1990s, when oil prices were record-low and global oil supply was extremely abundant, the studies of interest groups subsided (McFarland, 2004). In the 2000s, when oil prices sky-rocketed, the studies of domestic interest groups grew in its significance again, and currently, in 2014–2015, oil became cheap again, reducing the influence of oil and natural gas interests ('Stocks Crushed, Oil Slides', 2015).

In addition to studies focused on the energy sector, scholars explore regional and country-specific domestic groups. For example, Russia has been categorized as a typical case of the corporatist model, in which certain interest groups maintain the monopoly in specific issue-areas, as opposed to the pluralist model that implies the diversity and competition among interest groups (Tolstykh, 2006, pp. 11–14). The Russian government encourages a monopolistic domestic structure in order to establish a symbiotic relationship with interest groups by giving preferential treatment to some of them. While this structure was relaxed in the 1990s during privatization, which reduced the government monopoly over commodities, the structure has become increasingly monopolistic again under Vladimir Putin (Colton, 2008, p. 264).

In sum, domestic groups have powerful potential to influence states' energy policy decision-making via lobbying the lawmakers and persuading the public in different political systems and contexts.

An integrated framework for analyzing energy cooperation

From the brief literature review above, it is still clear that an enormous amount of research has been done in various areas pertinent to energy security. However, those were mutually exclusive theoretical approaches, and they each examine a particular set of cases that support their analysis. The literature is fragmented and isolated: fragmentation leads to a narrow understanding of energy security processes and stalemates in policy-making decisions. To broaden our understanding of energy security issues, this study offers a more integrated approach that considers multiple variables, borrowed from previous schools of thought, and applies them to several important case studies. This integrated approach allows examining an energy issue from several aspects in a systematic way, both for comparative analysis and for policy recommendations.

Previously, other scholars have attempted such an approach. For example, Peter Rutland suggests several models for explaining Russian energy policies in the Caspian region. First, the *kuwaitization model* argues that Russia's energy resources serve as the country's competitive advantage in the international division of labor. Second, the *market forces model* is based on liberalism and advocates free trade, privatization, and the prevalence of market mechanisms. Third, the *rent-seeking model* prioritizes rent-seeking over profit-seeking: rent-seeking is the extraction of benefits from distortions in the economic system (such as the preeminence of the energy sector). Fourth, the *'Russian bear' model* promotes the state's national interest, similar to realism. Finally, the *pluralist model* describes competition among rival groups in a society. Rutland compared a country's policies to a mosaic, which can be explained partly by universal economic theories, partly by geopolitics, and partly by the interests of the elite (Rutland, 2000). In other words, Rutland finds that even over a short decade in the 1990s business and politics continued shifting so much that no one model emerges as the best explanation of major energy policies of that period.

Another excellent volume, edited by Jeronim Perovic, Robert Orttung, and Andreas Wenger, offers a multi-aspect assessment of Russia's energy policies in Eurasia, including Russia–EU and Russia–U.S. relations, as well as Caspian states. Assessing the case of Russia from different dimensions, the book provides a wealth of knowledge on current Russian energy ties, but it does not have the goal of creating an integrated framework or a comprehensive literature review (Perovic, Orttung, & Wenger, 2009). The authors also conclude that the field needs a multi-variable theoretical model, rather than advancing a single explanation.

Another multidisciplinary approach to energy security has been offered in a recent volume by Gawdat Bahgat. The author explains the benefits of a multidisciplinary model. First, the energy mix in most countries is diversified and includes coal, oil, natural gas, nuclear energy, and renewables. Different sources of energy require different approaches. States may be self-sufficient in some aspects, such as coal, and highly dependent on foreign suppliers for others, such as oil. This will lead to different policy decisions in terms of how to ensure a stable supply of vital resources. Second, energy policies involve multiple aspects, such as investments, resource nationalism, geo-politics, and others. The diversity in these aspects requires studying appropriate literatures that generate knowledge on these particular issues. He concludes that energy security may only be explained by a combination of disciplines and areas of knowledge (Bahgat, 2011).

Inspired by these early efforts, this study offers an integrated framework that could simultaneously evaluate geopolitical, economic, and domestic politics variables across key cases. These variables have been refined to generate several hypotheses applied to the cases of U.S.–Russia, U.S.–Azerbaijan, and Russia–Germany. These energy relations unfold in the context of four key dimensions.

The international dimension

The international context refers to broad conditions in the international system, affecting energy trade and investments. Examples of such factors include global business and economic environments, world energy markets, oil prices, multilateral institutions, and capital movements (Razavi, 2007, pp. 187–188).

Over the last decade, there have been changes in the global patterns of oil supply and demand. According to the BP research, the biggest oil demand in the 1990s was observed in the Organization for Economic Co-operation and Development (OECD) countries, while OPEC countries provided most of the supply. Since 2000, however, oil demand has been growing faster in developing countries (e.g. China), rather than in the OECD. Similarly, oil supply has shifted from traditional sources (OPEC countries) to non-OPEC countries, including Russia and states in the Caspian region, such as Azerbaijan (Khalilov, 2009). Non-OPEC countries have grown in importance for global oil supply because of fewer limitations on their production volume and price-setting.

In order to monitor such trends in supply and demand, governments created international institutions: for example, the International Energy Agency (IEA), which includes 28 country-members, monitors energy supplies and provides fuel-security guidelines designed after the Middle Eastern oil embargos of the 1970s (Aalto & Westphal, 2008, p. 10). The IEA Chief Economist Fatih Birol predicted in 2011 that the 'era of cheap oil is over', because the booming demand for human mobility (automobiles and air travel) will continuously drive oil prices up. The number of oil producers in the world will remain limited. Since the number of oil-possessing states is lower than the number of natural gas-possessing countries, the importance of conventional and shale natural gas will continue to grow in the global energy, limited only by environmental concerns about shale gas production – water pollution, high water consumption, earthquakes, methane leaks, among others. In addition to regular natural gas delivered pipelines, liquefied natural gas (LNG) is becoming a globalized commodity, as LNG is now transported by tankers around the world (Tay, 2011).

Multinational institutions, such as IEA, guide the cooperation in global energy resources, helping mitigate risks of energy investments for individual countries. To reduce risks, the Multilateral Investment Guarantee Agency (MIGA), a part of the World Bank Group, provides 'investment guarantees against the risks of currency transfer, expropriation, war, civil disturbances, and breach of contract by the host government' (Razavi, 2007, p. 69). Azerbaijan and Russia are both MIGA-eligible countries ('MIGA Member Countries (173)', 2008). Other multilateral agencies that provide financial support and advice to governments include the European Investment Bank (EIB), the European Bank for Reconstruction and Development (EBRD), the OECD, the Asia Pacific Economic Cooperation (APEC) Energy Working Group, the North American Energy Working Group, the International Energy Forum, and the OPEC Fund for International Development. They work together with national agencies, such as the U.S. Agency for International Development (USAID) and the U.S. Export-Import Bank (USExim) (Razavi, 2007, pp. 105–107).

Multilateral agencies have been viewed as most effective in promoting sustainable development and energy independence around the world:

> In a world of increasing interdependence between net exporters of energy and net importers, it is widely recognized that multilateral rules can provide a more balanced and efficient framework for international cooperation than is offered by bilateral agreements alone or by non-legislative instruments.
>
> ('About the Charter', 2009)

On global issues, states join their efforts under several treaties: the Kyoto Protocol, administered by the 1992 United Nations Framework Convention on Climate Change (UNFCCC), greenhouse gases emission reductions, is an example of such a treaty in the area of climate change (Geri & McNabb, 2011). Another pivotal agreement on climate change has been negotiated in 2015 and will be reviewed at the Conference of the Parties (COP) to the UNFCCC in December 2015 in Paris, France. As Secretary-General of the United Nations correctly put it, '[a]ddressing climate change is essential for realizing sustainable development. If we fail to adequately address climate change, we will be unable to build a world that supports a life of dignity for all' ('2015 pivotal for finalizing universal climate change agreement, Ban tells Member States', 2015). By sustainable development, he meant 'low-carbon, climate resilient growth'.

International treaties are not easy to implement and enforce, especially when their guidelines and requirements contradict national energy policies. Also, the energy sector is usually a part of the economy in which nations prefer to keep as much sovereignty as possible. For example, Germany went ahead with the Nord Stream pipeline (discussed in Chapter 4) despite substantial EU opposition.

Despite the existence of multilateral guidelines, most decisions in energy cooperation are made bilaterally alongside such issues as asymmetry, mutual dependency, protectionism, state monopolies on energy infrastructure, and export restrictions (Barysch, 2005).

The national level

The state dimension includes a country's economic conditions, domestic legislation and regulations, as well as the overall business and political climate. States introduce regulations and restrictions that affect the volume and the nature of energy trade, balancing between free-market rules and government control.

Even developed economies do not always push for international openness for their energy industries and businesses. In the European Union, nations prefer to protect their national companies rather than allowing companies from other European countries to compete on their domestic markets (Aalto & Westphal, 2008). In the cases of the United States and Germany, both countries have carefully protected their energy markets from too much foreign influence. In 2005, for example, the U.S. government blocked the Unocal deal, when a Chinese company wanted to buy a major U.S. energy company. Former CIA chief R. James Woolsey even said that it was 'naive' to assume that the Chinese National Offshore Oil Corporation's bid for Unocal was 'just a commercial matter, unrelated to a Beijing strategy for domination of energy markets and the Western Pacific' (Grier, 2005). As a result, the U.S. Congress voted 398 to 15 to block the deal.

In addition to outright bans for companies' takeovers, protectionism includes other trade barriers and restrictions, as well as preferential conditions for domestic producers and consumers. For example, Russia has been criticized that its domestic natural gas prices have been much lower than world market prices. The Russian government has responded that Russia's domestic consumers would not be able to pay world prices and that a plentiful supply of Russian natural gas helps keep domestic prices down (Barysch, 2005, p. 25). Such protectionism is very much supported by Russian consumers, who favor Russia's four-tier pricing system for natural gas. Domestic consumers pay only about $75-$97 per thousand cubic meters of natural gas, compared to up to $450 that Europe pays to Russia for the same volume. Experts have noted that Russia sell natural gas to domestic consumers at a loss, because the cost of production is about $130 per thousand cubic meters ('Russia's Natural Gas Dilemma', 2012).

In addition to domestic prices, the prevailing state ownership of the energy infrastructure has been a debated issue ever since the 1970s, when a wave of energy sector nationalization occurred in the Middle East, Latin America, and other regions (described in great detail in Daniel Yergin's *Prize*). In Latin America, for example, we observe an ownership structure in the energy sector similar to the former Soviet regions: governments control major extractive industries and commodities (López-Morales & Vargas-Henández, 2014).

State ownership of energy industries is not a new concept in the global economy: gas pipelines, for example, 'have historically been undertaken by state entities' because of their long-term nature and large investments required (Razavi, 2007, p. 38). However, the government control has been an obstacle to cooperation between Russia and Western companies. The EU has also strongly opposed the state monopoly of Russian gas pipelines and strongly recommended that Russia open the energy sector to international competition. The Russian side did not accept these demands as realistic in the short term (Barysch, 2005, p. 25).

Relations between energy consumers and energy producers are often discussed in terms of *asymmetry*, which means that one side is more dependent on the other and is, therefore, vulnerable. An alternative situation is when dependency is mutual. For example, according to Katinka Barysch of the Centre for European Reform, EU–Russia gas trade is a case of mutual dependency, which inspired the EU–Russia Energy Dialogue in 2000 (Barysch, 2005, p. 27).

Finally, at the country level, states with a similar (relatively low) GDP per capita – such as Russia and Azerbaijan – have common economic problems and risks that are taken into account by decision-makers: limited domestic sources of capital, volatile legislatures, and unfavorable investment climates (Razavi, 2007, pp. 19–23). Such risks are measured and evaluated by different credit rating agencies (Moody, Fitch Ratings, Standard & Poor) and by multilateral institutions (Razavi, 2007, p. 162). One example of such evaluation is the 'Doing Business' reports published by the World Bank that rank countries according to the ease of doing business in those countries ('Doing Business 2009', 2008).

To sum up, the national dimension provides valuable insight into the conditions in each country that affect all industries (e.g., an investment climate, political stability), as well as specific economic aspects for the energy sector (e.g., proven resources of oil and gas, government controls, and other state policies), especially in countries with almost half of the state budget revenues provided by oil and gas exports.

The industry dimension

The importance of the oil and gas sector can be judged partially by its share in a country's GDP, the amount of oil and natural gas reserves in the country, the institutional structure of the industry, and the volume of domestic energy supply and demand (Razavi, 2007, p. 182). Along these indicators, Russia and Azerbaijan have been rather similar: the oil and gas industry is a major revenue generator for these states. Their energy production and revenues are discussed in detail in Chapters 2 and 3.

Energy production and trade are done by two types of businesses: major companies (both state-owned and private) and smaller independent enterprises. They have different preferences in energy-market operations and employ different market strategies while evaluating possible business opportunities for their involvement in energy projects. Major oil and gas companies usually prefer long-term investment with the large volume of operations, in order to increase the return on their investments. Since the 1970s, major oil companies have widely used futures contracts that obligate a seller and a buyer to implement a transaction at a fixed price on an agreed date. Such contracts add certainty to business operations. Oil futures are also used as an investment.

The opposite form of trading – the day-to-day spot trading – became widespread in the 1980s, when the liquidity of oil as a commodity increased dramatically. Spot trading means for the oil business that a 'cargo of oil is bought and sold many times, as a virtual asset, before oil physically reaches a refinery' (Razavi & Fesharaki, 1991, p. 3). Currently, major oil companies use both types of trading, relying mainly on futures contracts while using the spot market to fill gaps in supply. Smaller independent oil companies, in their turn, mostly operate on flexible spot markets (Razavi & Fesharaki, 1991, pp. 22–24). This study focuses on bigger projects such as pipelines and oil production that are normally implemented by major oil companies, rather than by smaller independents.

Major energy companies continually choose between long-term and short-term trade strategies. Companies react to growing operational costs of oil production and fluctuating oil prices, global demand and supply by adjusting their volume of production. Companies are 'becoming more adept at maintaining long-term business strategy in the face of short-term uncertainty stemming from oil price cycles' (Dittrick, 2009).

Unlike oil, which is a global commodity with volatile prices, natural gas trade usually involves large-scale pipelines that require substantial long-term investment in sophisticated equipment: natural gas has to be transported under pressure, with all the precautions against possible leaks and explosions (Blanchard, 2005, p. 14). In addition to technical challenges, multi-billion-dollar natural gas pipelines, once built, result in a high level of commitment and dependency between the supplier and the buyer (Wu et al., 2008, p. 2). This dependency is further complicated by geopolitical risks, such as the use of transit countries with unstable political systems and ethnic conflicts.

Natural gas shortage in many regions, especially in the United States, has been partially alleviated by the hydraulic fracturing technology that is utilized to extract shale natural gas and oil. The technology of hydraulic fracturing, first invented in the 1940s, now allows unlocking oil and natural gas in the shale, which was not technologically accessible or economically feasible before. Hydraulic fracturing (also known as fracking) involves drilling a well vertically, then horizontally, and injecting a mixture of water and chemicals in order to crack the shale rock and to let oil flow. According to oil and gas experts, the majority of new natural gas wells in the United States now use fracking (Warren & Nysveen, 2011). Globally, the unconventional natural gas resources (meaning, extracted by fracking technologies) are five times bigger than conventional reserves (i.e. extracted without fracking, by traditional methods) (Dittrick, 2011).

Shale gas has completely changed the situation on domestic gas markets. Before the 'shale revolution', the 2006 Annual Energy Outlook by the U.S.

Department of Energy predicted that by 2030 the United States would have to import 4.5 trillion cubic feet of natural gas per year. In 2010, this prediction was lowered to 1 trillion cubic feet per year. Before the American shale revolution, Russia was the biggest producer of gas in the world from 2002 to 2008.

In 2009, the United States became the biggest producer (624 billion cubic meters) of natural gas. At the same time, the United States has been the biggest consumer of natural gas (650 billion cubic meters per year) (Alexandrov, 2010). Already in 2013, the United States not only satisfied its domestic demand for natural gas through shale production, but it even became a natural gas exporter. Shale gas constituted more than one third of the overall natural gas production in the United States, while its shale production was only 3 per cent in 2004 ('United States: Shale Gas Boom to Boost Industry, Trade', 2013).

The United States and Canada, as well as Europe, possess abundant shale gas reserves. Major energy companies participate in large-scale shale projects in the United States, stirring a wave of mergers and acquisitions. Exxon Mobil purchased XTO, one of the largest American producers of shale gas. Statoil became a partner with Chesapeake in a joint venture in shale gas production in Markellus in Quebec. BP, Eni, and BG have also entered the American shale gas market (Alexandrov, 2010, pp. 60–64). As a result, natural gas production in the United States has soared, while the demand has been low because of the economic recession of 2009.

The industry has estimated that with the use of fracking, the United States has accessible natural gas reserves equal to about a century of natural gas supply. Fracking made the U.S. independent in terms of natural gas needs. Currently, the United States is working on expanding LNG terminals and export facilities ('Research and Markets: Global Liquefied Natural Gas (LNG) Market Assessment 2014–2025', 2015). Those terminals, however, will require billions of dollars of investments for all that infrastructure to be built. Ethane (a byproduct of petroleum refining) and propane exports are also predicted to rise (Nieto, 2015).

Fracking is as prevalent in oil production, as it is in gas production. The oil and gas analysts have estimated that the shale revolution in oil production has doubled the number of oil rigs in the United States, making the country one of the leaders in global oil production (Warren & Nysveen, 2011).

In Europe, the geology and dense population have so far prevented fast development of shale oil and gas extraction. According to Shell, Europe has about 1,200 trillion cubic feet of shale gas, and the biggest shale gas fields are Alum Shale (Sweden), Silutian Shale (Poland), and Mikulov Shale (Austria) (p. 66). However, shale gas is unlikely to drastically change the domestic natural gas supply in Europe in the near future. By 2010, European producers had drilled about 100 wells in Europe, as compared to more than 900 wells in the United States. In addition to the low number of gas wells, Europe is a densely populated continent, and shale gas production may create problems for the environment (Savelyev, 2010, p. 66).

Still, global oil and gas companies have kept an eye on shale opportunities in Germany, Hungary, Poland, England, and Romania (Dittrick, 2011). Western

Europe has not been too enthusiastic about shale production due to serious environmental concerns, but the recent Russia-Ukraine crisis might change that. The United Kingdom has already discussed tax incentives for shale production, and Denmark has started to issue shale permits ('Western Europe Oil and Gas Insight – April 2015', 2015, p. 4).

In the meantime, while the shale production has not taken off in those countries, their governments have evaluated the idea of importing liquefied natural gas from the United States. For example, Poland's Deputy Economy Minister Andrzej Dycha suggested that it might be cheaper for Poland to import natural gas from the United States than to rely on the current supply mix ('Gas is Most Important for Poland in TTIP', 2015).

As LNG is becoming a more economically feasible option, the global oil markets have also experienced significant changes. In 2014, oil prices unexpectedly dropped: they reached as low as $40 per barrel, compared with their highest point of about $140 per barrel a decade ago. The drop in oil prices has been explained by a number of factors: oversupply by OPEC, geopolitical factors, weak demand because of an economic recession, and shale oil revolution in the United States ('Global Outlook of the Energy & Environment Industry: Top 15 Trends', 2015).

The drop in oil prices has a direct implication for shale technologies. Oil extracted by fracking is more expensive than oil extracted by traditional methods. Oil companies might find fracking economically unfeasible if oil prices drop further, to the level of the 1990s (about $20 per barrel). Rex Tillerson, CEO of Exxon Mobil Corp, called the current drop in prices a resiliency test for the oil industry. Expressing the view of the whole industry, he remained optimistic, however, hoping that efficiency improvements would save the industry, and the output would remain high (Gilbert, 2015).

Economists have concluded that the United States overall will greatly benefit from low oil prices, which are a 'stimulus package' for the economy, stimulating domestic demand and increasing purchasing power ('Global Outlook of the Energy & Environment Industry: Top 15 Trends', 2015). American manufacturing industries, and especially petrochemicals, have indeed enjoyed lower production input expenses, because of cheap oil and natural gas ('House Science, Space, and Technology Subcommittee on Energy Hearing', 2015).

Other oil-producing countries, such as Russia and Venezuela, have not been so happy about low oil prices. Economists predict that those countries might slide into an economic recession, because lower oil revenues result in shortages in their federal budgets ('Global Outlook of the Energy & Environment Industry: Top 15 Trends', 2015).

The project level

All those industry conditions influence multiple individual projects, such as pipelines, offshore production sites, and refining plants. Each project is analyzed by investors (shareholders, governments, companies, bilateral sources, multilateral institutions, export-import banks) along several criteria. One group of criteria is *technical* and includes design, engineering, procurement, construction, safety, maintenance, environmental assessment, auxiliary services, and technical assistance. The other group is *financial* and evaluates potential expenses and revenues (Razavi, 2007, p. 207). Large projects usually have multiple investors: the diversity of investors makes a project more legitimate in the eyes of the population and provides an additional source of security for investors.

Depending on a company's risk tolerance, financing of a project can be *recourse* or *non-recourse*. In recourse projects, an operator owns a share of the whole corporation, not only of a specific energy project. Non-recourse financing means that a corporation is liable only within a specific energy project. In the latter case, the risk for outside investors is higher, because the rest of the corporation is not liable if the project fails. Usually both outside investors and project operators try to find the middle ground through limited-resource financing, preferred shares, and leasing (Razavi, 2007, pp. 6–9).

In countries such as Russia and Azerbaijan, the investment agreement is signed either with the government or a government-controlled enterprise. Large-scale oil and gas exploration projects in such countries often start with signing a production sharing agreement (PSA) that specifies a tax structure, licenses, environmental policies, profit sharing, and other conditions of a certain project. Participating parties take into account price forecasts, the size of reserves (proven, probable, or possible) and their productivity, taxes, royalties, a depletion premium (which is the opportunity cost of extracting the resources sooner rather than later), and environmental concerns, before signing a PSA (Razavi, 2007, pp. 217–229). Environmental and social concerns may delay the implementation of a project until the investors comply with regulations and provide all necessary evidence and documentation. Considerations include 'geology and geography, engineering, costs, investment, logistics, and the mastery of technological complexity' (Yergin, 2011, p. 49).

PSAs are designed to protect an investor in a volatile business environment in which tax structures and legislatures often change. However, in some cases (as happened in Russia), host governments may revise the PSA, if market conditions change not in the host country's favor. Smaller independent companies, unlike major corporations, usually resort to portfolio investments rather than PSAs, since the scope of their operations is narrower, and they want to be able to easily pull out of a host country in case of adversity (Johnson & Robinson, 2005, p. 189).

When companies evaluate energy projects to join, criteria developed by the World Bank and other multinational institutions help them mitigate their risks. In addition to information support, the World Bank and other institutions provide loans to companies and governments on various conditions. The World Bank's Energy Sector Management Assistance Program (ESMAP), for example, evaluates renewable energy projects and lends approximately $2–3 billion per year in those projects. ESMAP, together with the Global Environment Facility (GEF) and International Development Association (IDA), provides loans to low-income countries with GDP per capita of $1,065 or less on favorable terms with a low interest rate ('International Development Association: Borrowing Countries', 2009). Unlike the World Bank, which deals only with governments, the International Finance Corporation (IFC) provides loans directly to companies (Razavi, 2007, p. 65).

Cause and effect: explaining energy cooperation

The sections above outlined the complexity of energy cooperation on multiple levels. These will guide the explanation of the detailed cases under consideration: US–Russia, U.S.–Azerbaijan, and Russia–Germany. The cases will be structured to provide revealing tests of the variables inspired by prevailing theories of international relations (liberalism, realism, and domestic politics). The independent variables – economic potential, geopolitical rivalries, and domestic interest groups – will determine the dependent variable – energy trade and investment. Using these variables, several hypotheses have been generated for application to each bilateral case.

Economic potential hypothesis 1: If the economic potential of energy cooperation is high (i.e. the investment climate is favorable for foreign investors and energy resources are plentiful), the probability of cooperation will increase.

This basic hypothesis assumes that economic efficiency is the primary criterion for most projects and that corporations and governments try to maximize their profits. Investors make their decisions, based on a number of criteria: 'political stability, the tax regime (rates, transparency/predictability, and enforcement/appeals), and property rights (whether shareholder rights or intellectual property rights)' (Marshall, 2003, p. 11). These together form an investment climate.

Economic potential hypothesis 2. If economic potential of energy cooperation is very high, it will overcome geopolitical rivalries and other political differences.

In some cases, economic potential is very high. For example, geographic proximity and the ease of transportation by ocean routes make energy trade a natural fit, despite long-term hostilities between governments and differences in their domestic political systems. The economic potential of crude oil import-export also depends on the cost of production and the shipping costs of crude oil, since the price of oil itself is globalized.

Geopolitical hypothesis 1: The probability of energy cooperation is inversely related to the level of geopolitical rivalry between the two states.

The assumption here is that geopolitical rivalry impedes cooperation. The profitability of energy projects is not the first priority here, because a geopolitical rivalry prioritizes other considerations: domination in the region, the value of the land, access to strategic resources, and economic/resource independence. The projects chosen may or may not be most cost efficient: for example, a longer pipeline may be built instead of a shorter, cheaper one, just to avoid certain countries or territories.

Geopolitical hypothesis 2: The level of a government's geopolitical prioritization in energy policy is directly related to the level of state ownership in the country's energy sector.

As of 2009, out of the top ten oil and natural gas companies (in terms of the produced petroleum volume) in the world, seven were state-owned (Saudi Aramco, National Iranian Oil Company, Pemex, Petroleos de Venezuela, Kuwait Petroleum Company, PetroChina, and Sonotrach – listed in descending order). Only the remaining three companies (British Petroleum, ExxonMobil, and Shell) were private. If the companies are ranked according to their oil reserves, there is only one private company in the top ten list: Russia's LUKoil (Lee, 2009, p. 63). Four years later, in 2013, six out of the top ten oil and natural gas companies are state owned – in China, Iran, Mexico, Kuwait, Saudi Arabia, and Russia. The top five largest oil companies in the world in 2013 were PetroChina, National Iranian Oil Company, Saudi Aramco, ExxonMobil, and Gazprom (in random order) ('The Top 10 Biggest Oil & Gas Companies in the World Revealed and Profiled', 2013).

Domestic interest groups hypothesis 1: If powerful domestic groups benefit from (or oppose) oil/gas cooperation, the probability of cooperation will increase (or decrease).

The assumption behind this basic hypothesis is that domestic groups will be interested in maximizing their own power and influence, advancing their interests. If they see that energy cooperation helps to achieve these goals, they will support cooperation. Another assumption is that groups are, in fact, able to influence government policies, both in democratic and authoritarian states, as well as in transitional political systems (such as Russia and Azerbaijan).

Domestic interest groups hypothesis 2: If such interest groups are divided, the state's autonomy in pursuit of energy security via energy policy formations will increase.

For example, in the conflict between Azerbaijan and Armenia, the powerful Armenian lobby in the U.S. has been more concerned with the issue of Nagorno-Karabakh and with Turkish recognition of the 1915 genocide than with the issue of pipelines (Ismailzade, 2005). The Armenian Diaspora has not been focused on the issue of blocking Azerbaijan's (and promoting Armenia's) interests in energy deals, such as the BTC pipeline. The Armenian lobby, as well as a growing Azerbaijani lobby in the United States, promotes Armenia's and Azerbaijan's interests, respectively, especially on the issue of the secession conflict in Nagorno-Karabakh. At times, however, the Armenian Diaspora has also differed with the foreign policy priorities of Yerevan. Here, the Diaspora's politics is examined in relation to state policies.

These hypotheses provide a structural context for analyzing the bilateral cases of U.S.–Russia, U.S.–Azerbaijan, and Russia–Germany. This study employs the case method, based on economic, political, and industry data. Bilateral cases were selected because most energy deals, especially in the oil and natural gas areas, are still controlled and guided by governments. Each case study embraces the complexity of energy projects and takes into account relevant international and multilateral factors. A structured review of each dyad provides evidence for the theoretical predictions above, because case studies allow the exploration of details and options available to decision-makers in specific economic and political contexts.

References

2015 pivotal for finalizing universal climate change agreement, Ban tells Member States. (2015, February 24). *M2 Presswire*.

Aalto, P. (2008a). The EU-Russia Energy Dialogue and the Future of European Integration: From Economic to Politico-Normative Narratives. In P. Aalto (Ed.), *The EU-Russian Energy Dialogue: Europe's Future Energy Security* (pp. 23–41). Hampshire: Ashgate Publishing.

Aalto, P. (Ed.). (2008b). *The EU-Russian Energy Dialogue: Europe's Future Energy Security*. Hampshire: Ashgate Publishing.

Aalto, P., & Westphal, K. (2008). Introduction. In P. Aalto (Ed.), *The EU-Russian Energy Dialogue: Europe's Future Energy Security* (pp. 1–21). Hampshire: Ashgate Publishing.

About the Charter. (2009). Retrieved May 1, 2009, from http://www.encharter.org/index.php?id=7

Adams, N. (2003). *Terrorism and Oil*. Tulsa, OK: PennWell.

Alexandrov, V. (2010). V Slantsi Mi Verim. *Neft Rossii*, (6), 60–64.

Axelrod, R., & Keohane, R. O. (1985). *Achieving Cooperation under Anarchy: Strategies and Institutions*. Baltimore, MD: The Johns Hopkins University Press.

Bahgat, G. (2011). *Energy Security: In Interdisciplinary Approach*. Hoboken, NJ: Wiley.

Banks, F. (2007). *The Political Economy of World Energy: An Introductory Textbook*. Singapore: World Scientific Publishing Co.

Barysch, K. (2005). The EU Perspective. In D. Johnson & P. Robinson (Eds.), *Perspectives on EU-Russia Relations* (pp. 21–34). New York, NY: Routledge.

Baumgartner, F. R., & Leech, B. L. (1998). *Basic Interests: The Importance of Groups in Politics and in Political Science*. Princeton, NJ: Princeton University Press.

Blanchard, R. D. (2005). *The Future of Global Oil Production: Facts, Figures, Trends and Projections, by Region*. Jefferson, NC: McFarland & Company, Inc.

Bueno de Mesquita, B. (1997). A Decision Making Model: Its Structure and Form. *International Interactions*, 23: 235-266.

Carter, R. G. and Scott, J. M. (2004). Taking the Lead: Congressional Foreign Policy Entrepreneurs in U.S. Foreign Policy. *Politics & Policy*, 32: 34–70. doi: 10.1111/j.1747-1346.2004.tb00175.x

Closson, S. (2009). Russia's Key Customer: Europe. In J. Perovic, R. Orttung, & A. Wenger (Eds.), *Russian Energy Power and Foreign Relations: Implications for Conflict and Cooperation* (pp. 89–108). London: Routledge.

Colton, T. (2008). *Yeltsin: A Life*. New York: Basic Books.
Copeland, D. C. (1996). Economic Interdependence and War: A Theory of Trade Expectations. *International Security*, *20*(4), 5–41.
Dekmejian, R. H., & Simonian, H. H. (2003). *Troubled Waters: The Geopolitics of the Caspian Region*. New York, NY: I.B. Tauris.
Dittrick, P. (2009, May 5). OTC: Oil Firms More Adept at Long-term Strategies. *Oil and Gas Journal*.
Dittrick, P. (2011). Europe's Shale Revolution. *Oil & Gas Journal*, *109*(10A), 14.
Doing Business 2009. (2008) (pp. 1–211). Washington, DC: The International Bank for Reconstruction and Development/The World Bank.
EIA. (2012). Russia: Country Analysis Brief. Retrieved from www.eia.doe.gov
Falola, T., & Genova, A. (2005). *The Politics of the Global Oil Industry: An Introduction*. Westport, CT: Praeger Publishers.
Foyle, D., & Belle, D. V. (2010). Domestic Politics and Foreign Policy Analysis: Public Opinion, Elections, Interest Groups, and the Media. *The International Studies Encyclopedia* (Vol. II).
Gas is Most Important for Poland in TTIP. (2015, April 20). *Polish News Bulletin*, p. 4.
Geri, L., & McNabb, D. (2011). *Energy Policy in the U.S.: Politics, Challenges, and Prospects for Change*. New York: CRC Press.
Gilbert, D. (2015, April 21). Exxon CEO Expects Oil Prices to Remain Low in Coming Years; Rex Tillerson says price declines will test the shale revolution's resiliency. *Wall Street Journal (Online)*.
Gilpin, R. (2001). *Global Political Economy: Understanding the International Economic Order*. Princeton, NJ: Princeton University Press.
Global Outlook of the Energy & Environment Industry: Top 15 Trends. (2015, April 14). *PR Newswire*.
Goldman, M. (2008). *Petrostate: Putin, Power, and the New Russia*. Oxford: Oxford University Press.
Grieco, J., Powell, R., & Snidal, D. (1993). The Relative-Gains Problem for International Cooperation. *The American Political Science Review*, *87*(3), 727–743.
Grier, P. (2005, July 18). Unocal Deal Tests US Stance Toward China; Some Lawmakers Believe Chinese Ownership of the Oil Firm would Damage National Security. Others See No Harm. *The Christian Science Monitor*, p. 2.
Grossman, G. M. and Helpman, E. (2002). Integration versus Outsourcing in Industry Equilibrium. *Quarterly Journal of Economics*, February 2002, 117(1), pp. 85–120.
House Science, Space, and Technology Subcommittee on Energy Hearing. (2015). Lanham, MD: Federal Information & News Dispatch, Inc.
International Development Association: Borrowing Countries. (2009). Online: International Development Association (IDA).
Ismailzade, F. (2005). The Geopolitics of the Nagorno-Karabakh Conflict. *Global Dialogue*, *7*(3/4), 104–111.
Johnson, D., & Robinson, P. (2005). Perspectives on EU-Russia Relations. New York:, NY: Routledge.
Kaldor, M., Karl, T. L., & Said, Y. (Eds.). (2007). *Oil Wars*. Ann Arbor, MI: Pluto Press.
Keohane, R. O., & Nye, J. S. (1989). *Power and Interdependence* (2nd ed.). Glenview, IL: Pearson Scott Foresman.
Khalilov, S. (2009, July 5). Lecture to the Energy Politics and Economis Summer School Participants.
Krasner, S. D. (1978). *Defending the National Interest*. Princeton, NJ: Princeton University Press.

Lee, H. (2009). Oil Security and the Transportation Sector. In K. S. Gallagher (Ed.), *Acting in Time on Energy Policy* (pp. 56–88). Washington, DC: Brookings Institution Press.

López-Morales, J. S., & Vargas-Henández, J. G. (2014). Effect of the Type of Ownership in the Financial Performance: The Case of Firms in Latin America. *International Business Research*, *7*(10), 125–132.

Lowery, D., & Brasher, H. (2004). *Organized Interests and American Government*. New York, NY: McGraw-Hill.

Marshall, Z. B. (2003). Russia's Investment Outlook. In J. H. Kalicki & E. K. Lawson (Eds.), *Russian-Eurasian Renaissance? U.S. Trade and Investment in Russia and Eurasia* (pp. 9–28). Washington, DC: Woodrow Wilson Center Press.

McFarland, A. S. (2004). The Energy Lobby. In C. S. Thomas (Ed.), *Research Guide to U.S. and International Interest Groups* (pp. 206–208). Westpoint, CT: Praeger.

MIGA Member Countries (173). (2008, October 8). Retrieved April 11, 2009, from http://www.miga.org/quickref/index_sv.cfm?stid=1577

Morgenthau, H. (1948). *Politics Among Nations: The Struggle for Power and Peace*. New York, NY: Alfred A. Knopf.

Nieto, F. (2015). U.S. Ethane Ready To Go Global. *Midstream Business*, 5, 66–70.

Olson, M. (1965). The Logic of Collective Action. Cambridge, MA: Harvard University Press.

Perovic, J., Orttung, R., & Wenger, A. (Eds.). (2009). *Introduction: Russian Energy Power, Domestic and International Dimensions*. London: Routledge.

Pleines, H. (2009). Developing Russia's Oil and Gas Industry: what Role for the State? In J. Perovic, R. Orttung, & A. Wenger (Eds.), *Russian Energy Power and Foreign Relations: Implications for Conflict and Cooperation* (pp. 71–86). London: Routledge.

Razavi, H. (2007). *Financing Energy Projects in Developing Countries*. Tulsa, OK: Penn Well.

Razavi, H., & Fesharaki, F. (1991). Fundamentals of Petroleum Trading. New York, NY: Praeger.

Research and Markets: Global Liquefied Natural Gas (LNG) Market Assessment 2014–2025. (2015, April 7). *Business Wire*.

Rubenzer, T. (2008). Ethnic minority interest group attributes and US foreign policy influence: A qualitative comparative analysis. *Foreign Policy Analysis*, 4(2): 169–185.

Russia's Natural Gas Dilemma. (2012). *EurActiv.Com*, (April 11). Retrieved from http://www.euractiv.com/energy/russias-natural-gas-dilemma-analysis-512092

Rutland, P. (2000). Paradigms for Russian Policy in the Caspian Region. In R. Ebel & R. Menon (Eds.), *Energy and Conflict in Central Asia and the Caucasus* (pp. 163–188). Lanham, MD: Rowman and Littlefield.

Sachs, J. D., & Warner, A. M. (1995). Natural Resource Abundance and Economic Growth. Washington DC: National Bureau of Economic Research.

Savelyev, K. (2010). Toplivo Buduschego Ili Product PR? *Neft Rossii*, (6), 65–67.

Schattschneider, E. E. (1960). *The Semisovereign People: A realist's view of democracy in America*. New Haven: Yale University Press.

Socor, V. (2008, March 8). South Stream Gas Project Defeating Nabucco by Default. Retrieved November 12, 2008, from www.jamestown.org/edm/article.php?article_id=2372856 http://www.jamestown.org/edm/article.php?article_id=2372856

Standlea, D. (2006). *Oil, Globalization, and the War for the Arctic Refuge*. Albany, NY: State University of New York Press.

Stern, J. (2005). *The Future of Russian Gas and Gazprom*. Oxford: Oxford University Press.

Stiglitz, J. (2002). *Globalization and Its Discontents*. New York, NY: W.W. Norton and Company.

Stocks Crushed, Oil Slides; Oil Rig Count At 5-Year Low; More Big Oil Layoffs; Dow In The Red For 2015; Global Worries Hit Markets. (2015, April 17). *Finance Wire*.

Tanner, F. (2009, July 7). Energy as a Component of International Security. Lecture to Energy Politics and Economics Summer School Participants.

Tay, E. (2011). IEA Chief Economist Offers a Look at Our Energy Future. *Lowcarbonsg.com*. Retrieved from http://www.lowcarbonsg.com/2011/07/18/iea-chief-economist-offers-a-look-at-our-energy-future/

Tolstykh, P. A. (2006). *Praktika Lobbisma v Gosudarstvennoi Dume Federalnogo Sobrania Rossiiskoi Federatsii* [Lobbying of the Russian State Duma]. Moscow: Kanon.

The Top 10 Biggest Oil & Gas Companies in the World Revealed and Profiled. (2013, May 15). *PR Newswire*.

Truman, D. (1951). *The Governmental Process: Political Interests and Public Opinion*. New York, NY: Alfred Knopf.

United States: Shale Gas Boom to Boost Industry, Trade. (2013). Oxford: Oxford Analytica Ltd.

Warren, M., & Nysveen, P. M. (2011). The Shale Revolution Marches On. *Oil & Gas Investor, 31*(10), 1.

Western Europe Oil and Gas Insight – April 2015. (2015) (pp. 1–15). London: Business Monitor International.

Wolpe, B. C., & Levine, B. J. (1996). Lobbying Congress: How the System Works (2nd ed.). Washington DC: Congressional Quarterly Inc.

Wu, K., Fesharaki, F., Hosoe, T., & Isaak, D. (2008). Strategic Framework for Energy Security in APEC 2008. Retrieved April 8, 2009, from http://www.ncapec.org/reports/EWC_APEC%20Framework%20Report.pdf

Yergin, D. (1992). *The Prize: The Epic Quest for Oil, Money, and Power*. New York, NY: Simon & Schuster.

Yergin, D. (2011). *The Quest: Energy, Security, and the Remaking of the Modern World*. New York: Penguin Books.

2 U.S.–Russia energy ties

Introduction

U.S.–Russia economic and security relations have seen extreme ups and downs over the last two decades. In the 1990s, Western investments were flowing to Russia, and the Clinton administration was actively advising Boris Yeltsin's government on how to transition to the market economy. Encouraged by Western governments, the largest energy companies (such as Chevron, ConocoPhillips, ExxonMobil, and Shell) signed production-sharing agreements and started to invest billions of dollars in Russia's energy sector, hoping for quick profits. At the time, Russia's democratization and market reforms were praised by the West, despite the turmoil, corruption, and the overall unpredictability of the Russian economy. U.S.–Russia relations stayed productive even during the crisis of 1998, when the ruble collapsed and the country defaulted on its debts.

The economic chaos and domestic instability in the 1990s, aggravated by extremely low oil prices, were the starting conditions for Vladimir Putin when he was appointed prime minister by Boris Yeltsin and later elected as president. Putin began to reform the extremely inefficient taxation system and promised voters to establish the rule of law. Initially, these reforms paid off and stabilized the economy, bringing inflation down. Leading oligarchs had to either comply with Putin's demands or leave the country. In foreign affairs, Vladimir Putin and George W. Bush started their relationship on a relatively positive note, especially when Russia offered condolences and assistance to the United States after September 11, 2001.

Western companies were coming to Russia to invest in many areas of the economy, including large oil and natural gas exploration projects. The growth of U.S.–Russia trade and investment during this period may be explained by high economic potential (the companies were making record profits) and low geopolitical tensions. The U.S. government strongly encouraged American companies to do business in Russia, providing them with informational support and financial incentives. On the Russian side, the country needed foreign capital and advanced technology to revive its oil and gas industries, among others.

Since the early 2000s, as global oil prices started to rise and Russia substantially increased its oil and gas revenues, the Russian government became more assertive internationally and confident in its relations with the West. As Vladimir Putin wanted to restore Russia's great power status, U.S.–Russia relations quickly turned into mutual distrust and hostility over such issues as the North Atlantic Treaty Organization's (NATO) expansion, the Iraq war of 2003, and the U.S. missile defense in Europe. During that decade, the tensions between Russia and the United States culminated during the Russia–Georgia War of 2008 over the breakaway region of South Ossetia. During that military conflict, the United States strongly supported Georgia in official statements and advising.

These intergovernmental tensions led to decreased American access to new energy projects in Russia. Domestically, the Russian government was increasing the state control over the energy sector, especially decision-making in energy deals with foreign companies. Some of the energy contracts that had been initiated in the 1990s were revised or even cancelled. The geopolitical rivalries were hurting energy cooperation.

U.S.–Russia relations substantially improved during Dmitry Medvedev's presidency in 2008–2012 (with Vladimir Putin as prime minister), when the newly elected U.S. President Barack Obama initiated the U.S.–Russia 'reset', offering a fresh start in 2009, after the lowest point in 2008. During the reset, American companies again received an opportunity to join Russia's oil and natural gas projects, signing several multi-billion deals in such regions as the Arctic and the Black Sea. However, since then, Western companies were allowed to invest mostly in the most challenging climactic areas in the Arctic and East Siberia, in which Russian companies did not have enough expertise.

In 2012, the international community was watching Russian presidential elections, as Vladimir Putin was reelected to presidency for his third term. This time, the presidential term was six (instead of four) years, thanks to a change in the Russian Constitution that had been implemented by Dmitry Medvedev. With Vladimir Putin in power as president again, the U.S.–Russia relations quickly deteriorated and returned to the pre-'reset' levels.

Both Russia and the United States implemented laws against each other. For example, Russia passed an adoption ban, prohibiting American citizens from adopting Russian children. The United States passed the Magnitsky Act, entitled after the late Sergei Magnitsky – an auditor who had tried to investigate corruption in Russia and later died in Russian custody. The Act prohibited those who the U.S. government believed were responsible for Magnitsky's death from entering the United States and using the American banking system.

In such unfriendly business environment, as Russia's domestic political system has strengthened the presidential power (Colton, 2008), Western energy companies have been forced to limit their operations in Russia or even sell their stakes in Russian companies.

Finally, the 2014–2015 events in Ukraine brought the U.S.–Russia relations to an even lower level. In response to Russia's annexation of Crimea and the events in Eastern Ukraine, when separatists in Donetsk and Luhansk publicly announced their pro-Russia orientation, the United States, as well as other Western states, announced economic sanctions on Russia, while Russia implemented its own sanctions on some Western goods. The events in Ukraine are addressed in detail in Chapter 5, while the current Chapter 2 provides a thorough overview of U.S.–Russia relations in the larger context.

The Russian energy sector and the U.S. involvement

Russia's economy has for a long time been dependent on the oil and gas industry. Several decades ago, by 1975, the Soviet Union became the biggest producer of oil in the world, at the level of 12 million barrels of oil per day. Oil and gas were mostly exported to Europe via 50,000 km of oil pipelines and 150,000 km of natural gas pipelines (the Baltic Pipeline System, the North-Western Pipeline System, Tengiz-Novorossiysk, Baku-Novorossiysk, Druzhba, and other pipelines). The 2,300-mile, 1.4 million bbl/d Druzhba pipeline was the largest among them, delivering oil to Europe (EIA, 2010). Oil and gas exports were the biggest source of revenue for the Soviet Union and the way to influence the Eastern European countries.

For several decades, the Soviet Union (and then Russia) and the United States were the main rivals in terms of the largest oil production. The U.S. oil production used to exceed Russia's until the mid-1970s at the steadily growing level of more than 5 million barrels a day. After 1974–1975, and until 1990, the Soviet oil production took off, and the Soviet Union started to lead with more than 10 million barrels per day, while the U.S. production started to slowly decline, until the shale oil revolution in the late 2000s (Gaddy, 2006). At the present moment, Russia and the United States, as well as Saudi Arabia, are still global leaders in oil production. Russia and the United States have reduced the role of OPEC by flooding the world market with plentiful oil (Krauss, 2015).

After the collapse of the Soviet Union, Russia inherited the majority of Soviet pipelines, but the political and economic chaos of the early 1990s led to a sharp decline in Russia's oil production (Goldman, 2008, p. 11). The oil and gas industry, outdated and lacking modern technologies, survived in the 1990s only by using the existing Soviet infrastructure. Newly independent Russia was struggling to keep oil production going, amidst economic 'shock-therapy' reforms, sweeping privatization, asset stripping, and extremely low oil prices (around $20 per barrel, compared to $140 a decade later). The industry badly needed Western expertise, financial resources, and technology: the Yeltsin administration invited Western investors and foreign advisors with open arms. The fact that Boris Yeltsin had 'mutual admiration' with the U.S. Ambassador Robert Strauss in 1991–1992 also helped to build the relationship based on trust. During the Clinton administration, American economic and technical assistance to Russia exceeded $2 billion (Colton, 2008, pp. 267–268).

As Russia was experiencing the enormous state budget deficit while the taxation and manufacturing industries were almost collapsed, oil and gas remained the main means of survival for the modernizing state. Oil and natural gas were what Russia could offer to the world market in abundance. Slowly, incoming Western investments and rising oil prices helped Russia increase its oil production volume by the end of the 2000s. Russia produced around 8 million barrels of oil per day in 1992, and that volume dropped to 6 million barrels in 1996, staying at that low level until 1998. After the financial and economic crisis of 1998, oil production started to steadily grow, accompanied by the growth in world oil prices, and reached 10 million barrels per day in 2007, which was almost the volume that the Soviet Union produced at its peak in the 1980s (Goldman, 2008).

Initially, in the 1990s, the oil industry in Russia needed to be completely reorganized, creating multiple new entities. In September 1991, the Ministry of Fuel and Energy was transformed into the Ministry of the Petroleum Industry (with Vagit Alekperov in charge). The Ministry established the Rosneftegaz energy company that was later divided into several independent entities. In November 1991, another major company, LUKoil, was formed by combining assets located in the Langepaz, Urengoi, and Kogalym oil fields. A year later, Rosneftegaz was transformed into Rosneft. In 1993, Yukos and Surgutneftegaz came into existence (Goldman, 2008, p. 61).

Many future Russian oligarchs managed to become owners of major oil and gas enterprises for an extremely low price. In 1995, the Russian government badly needed financial resources and started the Loans for Shares (LFS) initiative, under which national resource companies were auctioned off for a small fraction of their costs to banks that had lent money to the government (Frye, 2000, p. 4).

Simultaneously, foreign capital was flowing to Russia though dozens of large-scale deals and production-sharing agreements: ARCO bought $250 million worth of shares of Russia's LUKoil, receiving access to Siberian oil reserves (Yergin, 2011, p. 29). Total (France) invested $800 million in a project to develop the Kharyaga field in Timan-Pechora and to produce 30,000 barrels of oil per day. Royal Dutch Shell and ExxonMobil invested over $20 billion in oil and gas production in Sakhalin (Goldman, 2008, p. 63). In 1997, Mikhail Fridman's Alfa Bank and the U.S.-based Access Industries (chaired by Leonard Blavatnik), acquired 40 per cent of the Tyumen Oil Company. Vladimir Potanin's Sidanko established a partnership with BP and bought the Cleveland-based nickel producer QM Group, as well as the Montana-based palladium and platinum producer Stillwater Mining. Amoco (USA) bought a ten per cent stake of Sidanco, which later was sold to BP. Another major Russian oil company, TNK, signed a 198-million dollar contract for oil services with Halliburton, through a loan from the U.S. ExIm Bank. American service company Schlumberger started to work with Rosneft in the Yugansneftegaz Priobskaia oil field (Goldman, 2008, pp. 66–79).

In the 1990s, Russian companies did not have enough capital to invest in new production and struggled to maintain existing facilities (Hueper, 2003, p. 177). Russia needed investment from the West, as well as the cutting-edge technology that Russia did not possess at the time. As a result, by the mid-1990s, major international companies signed PSAs, creating more than one hundred joint ventures in the Russian energy sector. A standard PSA serves as a guarantee that the investing company returns its expenses. A PSA is a contract for 20–40 years, under which a foreign company provides capital, and the host government receives its share of profits only after the company returns its capital expenses (Blanchard, 2005, p. 205).

During that decade, the United States was one of the most important economic partners for Russia: American investments were steadily growing through the 1990s (Frye, 2000, p. 200). In 1995, investments from the United States comprised almost $1 billion and jumped to $3 billion in 1997. Even during the 1998 financial crisis, the American investments were still more than $2 billion, returning to $3 billion in 1999. Between 1995 and 1999, American investments accounted for up to half of all foreign investments in Russia, and most of it was in the energy sector (Arai, 2012).

Investment in Russia was considered safe, as before 2001 the political risks in the country were ranked as low. International companies were expanding their staff in Russia, receiving record profits, and praising tax and infrastructure reforms started by the newly appointed prime minister Vladimir Putin in 1999 (Marshall, 2003, p. 10).

As Russia's oil production was recovering after the economic turmoil of the 1990s, Russia even started to export oil to the United States: exports jumped from zero barrels per day in 1996 to more than 500,000 barrels per day in 2009, according to the Energy Information Administration. Moreover, by 2010, the United States was the seventh largest importer of Russian crude oil (at about 250,000–300,000 barrels per day), after Germany, the Netherlands, Poland, China, France, and Italy (EIA, 2010, p. 4). In the U.S. mix of suppliers, Russia's share was 3 per cent, which was the same as Iraq's, Nigeria's, and Angola's. For comparison, the United States imported 12 per cent of its oil from Canada, 7 per cent from Saudi Arabia, 7 per cent from Venezuela, 6 per cent from Mexico, and 4 per cent from Algeria (Dougher, 2009).

Russia's foreign policy was very West-oriented. In the beginning of their presidential terms in the early 2000s, Presidents Vladimir Putin and George W. Bush started the Russian–American Business Dialogue, promoted by such business groups as the U.S.–Russia Business Council (USRBC), the American Chamber of Commerce (AmCham), the Russian–American Business Council (RABC), and the Russian Union of Industrialists and Entrepreneurs:

> a business-to-business mechanism intended to expand contact between the two private sectors; to identify areas where laws, regulations, and practices impede trade and investment; and to provide a forum where business interests can be raised with the respective governments…
>
> ('Russian–American Business Dialogue', 2003)

Simultaneously, some Russian companies attempted to enter the American market. The most notable example was LUKoil, which in 2003 bought a number of gas stations from Getty Petroleum Marketing Limited on the U.S. East coast. The opening ceremony of a LUKoil gas station in New York was even attended by Russian President Vladimir Putin and U.S. Senator Charles Schumer (Maass, 2009, p. 185). In addition to its American assets, LUKoil purchased 376 Jetts gas stations in Europe from ConocoPhillips. Back in Russia, in 2004, ConocoPhillips bought 20 per cent of LUKoil stock for $7 billion (Goldman, 2008, p. 126).

The extreme closeness and friendliness between Russia and the United States started to deteriorate in the mid-2000s, when oil prices began to rise and the Russian economy oil income was rapidly growing. Russian energy companies enjoyed increasing revenues, and the need for foreign capital became less urgent (Hueper, 2003, p. 178). Having overcome the economic chaos of the 1990s and accumulating substantial oil revenues, Russia became much more assertive in geopolitical issues.

Simultaneously, Vladimir Putin started to strengthen his domestic power and state control over energy resources (Frye, 2010). The access by foreign energy companies to Russian oil and gas contracts was becoming more and more limited. As the Russian government realized that the PSAs signed with international companies in the 1990s gave a large share of revenues and power to foreign entities, it started to revise the existing PSAs and modified policies for signing new contracts. High oil prices meant that international companies received vast amounts of revenues that could instead go to Russia's state budget if it were not for the PSAs. Vladimir Putin even called the PSAs 'a colonial treaty', adding that those Russian officials who had signed them should be 'put in prison' (Goldman, 2008, p. 86).

As the Russian government was limiting access for most foreign entities, American companies especially felt those limitations. According to the Foreign Investment Advisory Council to the Russian government (FIAC), in 2004–2008, the overall share of American foreign investments in Russia (in all sectors of the economy) experienced the biggest drop among other investing countries (Johansson, 2008).

The relations between the United States and Russia experienced tensions (addressed in detail in the next section of this chapter). As a result, American companies were excluded from major oil and gas deals. According to the Russian Federal State Statistics Service, annual U.S. investments in Russia's energy sector comprised about $400,000 in 2005 and 2006, $570,000 in 2007, $1 million in 2008, and only $16,000 in 2009. Simultaneously, total foreign investments in the energy sector were around $5.1 million in 2005, $7.7 million in 2006, $16 million in 2007, $10 million in 2008, and $8 million in 2009. Most of the investments came from the Netherlands: $3.8 million in 2005, $3.6 million in 2006, $12 million in 2007, $4.8 million in 2008, and $2.7 million in 2009. These numbers show that American companies were especially disadvantaged, because the interests in investing in Russia remained high (FSGS, 2015).

An illustrative example of how foreign involvement in an oil and gas project was diminishing over time is the Sakhalin Island. Projects on the island were signed and implemented in several stages, and each stage received its own name, such as Sakhalin-1, Sakhalin-2, and so on.

Sakhalin-1 (for the Chayvo and Odupto oil fields) involved such companies as Exxon Neftegaz Ltd (USA) – 30 per cent of the project; SODECO Ltd. (Japan) – 30 per cent; ONGC Videsh Ltd (India) – 20 per cent; SMNG-Shelf (Russia)

– 11 per cent, and Rosneft-Astra (Russia) – 8.5 per cent. First oil production started in October 2005 (RussiaProfile.org, 2009).

The next project on the island – Sakhalin-2, covering the Piltun-Astokhskoye (oil) and Lunskoye (gas) fields – is managed by Sakhalin Energy Investment Co. Initially, the $20-billion Sakhalin-2 did not include Russian companies at all. However, in 2006, foreign participants were charged with violations of Russian environmental law, and Russia's Gazprom became the project's majority shareholder (Yergin, 2011, pp. 35–40).The participating parties currently include Gazprom (Russia) – 50 per cent of the project's capital; Royal Dutch Shell (the Netherlands) – 25 per cent; Mitsui (Japan) – 12.5 per cent; and Mitsubishi (Japan) – 12.5 per cent. The first oil was produced as early as 1999, while natural gas production started around 2007–2008. This project also included a LNG plant (RussiaProfile.org, 2009).

Sakhalin-3 had several parts to it, depending on what oil and gas field was explored. The Krinskii field exploration includes such companies as ExxonMobil (USA) – 33.3 per cent, Chevron Texaco (USA) – 33.3 per cent, and Rosneft (Russia) – 22.2 per cent. However, the project was suspended in 2007 because of the changes in the companies' exploration rights. A similar change happened in developing the Vostochno Odoptinski field of Sakhalin-3, in which ExxonMobil had 66.6 per cent of the capital investments and Rosneft had 33.3 per cent, and the project was suspended (RussiaProfile.org, 2009).

Other projects – Sakhalin-3 (for the Veninski field), Sakhalin-4 (for the Astrakhanovskii Offshore field), Sakhalin -5 (Vostochno-Schmidtovski, Kaigan/Vasukan, and Zapadno- Schmidovski fields), and Sakhalin-6 (for the Pogranichnii field) – do not include American partners at all, while Russian companies Rosneft and Gaspromneft have more than 50 per cent of control. Among foreign partners, some of the projects include Sinopec (China) and BP (UK) 49 per cent (RussiaProfile.org, 2009).

As such, revising contracts means imperfect intellectual property rights protection, complicated by Russia's underdeveloped financial and judicial systems and high bureaucratic obstacles to business. Excessive protectionism and discriminatory provisions in trade laws delayed Russia's accession to the World Trade Organization (WTO) ('Russian–American Business Dialogue', 2003). Russia was able to join WTO after an 18-year delay on December 16, 2012. By accepting Russia into the WTO, the organization is hoping that Russia will be able 'to tackle long-delayed economic reforms and to provide an improved regulatory and legal framework for investors, both foreign and domestic' (Iskyan, 2012).

Experts point out though, that the WTO delay has been not the only obstacle in normalizing the trade relations between Russia and the United States, beyond the U.S. sanctions after the Crimea annexation in 2014. Other laws, such as the Jackson-Vanik law, limit Russia's trade in the United States (Donohue & Shokhin, 2008). The Jackson-Vanik law is an American law that was passed during the Soviet Union in response to the USSR's poor immigration policies. Nevertheless, the law was valid until recently, preventing Russia from receiving the permanent normal trade relations (PNTR) status with the United States (Iskyan, 2012). The PNTR status was granted to Russia by the 112th U.S. Congress only in 2012 (Cooper, 2014).

In addition to traditional trade barriers, another way to limit access to oil contracts to foreign companies is to grant major oil fields the status of strategic importance, which has been done in Russia multiple times. In addition to specific oil and gas fields, the whole energy sector has been defined by Russian Ministry of Energy's official energy strategy for 2010–2030 as the 'most strategically important industry in the Russian economy'. The energy sector, the strategy stipulates, 'consolidates the regions of the Russian Federation and determines main economic indicators' (Ivanitskaya, 2009, p. 11).

To sum up, Russia inherited an extensive system of Soviet pipeline and the aging oil and gas sector that needs renovations, new oil wells, and massive investment. Yeltsin's domestic reforms of the 1990s (privatization, the loans-for-shares program, democratization, and market mechanisms) were plagued by high unemployment, inflation, corruption, and overall economic turmoil. Nevertheless, it was an optimistic time for American investments in Russia, which comprised about half of all investments coming to the country. The 1990s was also a period when major Russian private oil and gas companies signed multiple production sharing agreements and other joint projects with foreign energy companies. The economic crisis of 1998 helped Vladimir Putin be elected as president in 2000, as Russia was in search of stability and the rule of law promised by Putin. The U.S. investment in Russia was substantial until the beginning of the 2000s, when Vladimir Putin and George Bush started an energy dialogue. However, as Russia was gaining strength economically and geopolitically because of growing oil revenues, the Russian government started to disagree with the United States more often, which affected the business opportunities for American companies. The door for American businesses was practically shut because of such issues in the mid-2000s as NATO expansion, Kosovo, the Iraq war, and others. The share of American investments in Russia plummeted, especially in the energy sector. Some business opportunities were temporarily reopened during the Dmitry Medvedev–Barak Obama 'reset' but still strictly under the Russian government's control – geopolitical considerations were affecting economic calculations.

Economic potential for U.S.–Russia energy trade and investments

The opportunities for cooperation in trade and investment in the oil and gas industry depend first and foremost on economic factors: the oil and gas reserves in a host country and the prevailing investment climate. The oil and gas reserves, their methods of production and transportation also largely depend on a host country's geographic position.

Energy resource availability and accessibility

Oil

Russia possesses enormous oil and natural gas resources, although some of them are located in the areas that are difficult to access. Some of the largest oil fields in the world (Samotlor, Romashkino, Tevlin-Russkin, Vatyegan, Lyantor, Fedorov, Mamontov, Sutormin, and Povkhov, among others) are located in the East and West Siberia regions, where a harsh climate and remoteness make oil extraction and transportation especially costly and require significant investment (Blanchard, 2005). Since many of Russia's oil fields have been in use for several decades (some of them even go back to the 1950s), the oil industry needs to constantly discover new ones, in order to maintain the production volume constant (Gimbel, 2009, p. 92). Industry experts admit that the substitution of depleted wells with new oil discoveries has not been fast enough. Also, as older wells are depleted, new discoveries happen in ever more challenging areas such as East Siberia.

To deliver oil to the world market, oil companies transport it from in-land fields to seaports by pipelines and then export it by tankers. Because of the landlocked geographical position of Siberia and its remoteness from year-round navigational waters (the Arctic waterways freeze for several months a year, closing this transportation route), it has not been easy to deliver Siberian oil to international markets, and Russia has had to rely on transit states (the Baltic States, Ukraine, etc.) and to pay transit fees.

Through the 1990s, a major part of Russia's oil imports was directed toward Europe via the Latvian port Ventspils. Because Russia and Latvia could not agree on the conditions for the port use, in 2001, Russia built a port on the Russian territory – Primorsk – and it became the largest oil terminal in the Baltic region, allowing Russia to circumvent the problematic Latvian port. Primorsk exports 70 million tons (490 million barrels) of oil per year – twice the volume exported by another major Russian port – Novorossiysk (43 million tons, or 301 million barrels). Out of the total 1,400 million barrels of oil, which Russia exports annually, these two ports handle as much as 35 per cent and 20 per cent of the export volume respectively, without dependency on transit countries (Reuters, 2011). However, these two ports, together with other Black Sea ports, have been mostly relevant for exports to Europe. They are not as convenient for exporting oil to the United States and Asian markets.

It is a much more reasonable solution to deliver oil from the Russian East coast to the West Coast of the United States and to the Asian markets. Alaska already has all available infrastructure for accepting oil deliveries, and California has multiple refineries. However, for a long time, no pipeline existed to deliver oil to the Russian Far East.

Only a decade ago, the Russian government started to focus on the Pacific Ocean basin as the key destination for future energy export routes. To make such exports possible, the Russian government built a new pipeline from Siberia to the Pacific Ocean – the East Siberia-Pacific Ocean (ESPO) – with the planned capacity of 1.6 million barrels per day. It connects East Siberia with the Kozmino terminal (near Nakhodka, Primorsky Krai) on the coast of the Sea of Japan (the Pacific Ocean). The pipeline is scheduled to be built in two phases. The first phase for transporting 600,000 barrels per day was completed in late 2009, ending in Skovorodino, Amur Oblast, on the border with China. The pipeline was unveiled by Vladimir Putin, then prime minister of Russia, in 2009 and was praised as increasing export opportunities for Russia in the Pacific Rim. In China, a new refinery – the Daqing – was built in order to process the incoming Russian oil. The second phase of the ESPO pipeline, with the capacity of 1 million barrels of oil per day, is scheduled to open in 2015 – ending in Kozmino (Watkins, 2009).

The construction of the pipeline was followed by signing oil export contracts with Asian countries, such as Vietnam, which was a very typical example of Russia's current turn to Asia. The first shipment of 700,000 barrels of crude oil to Vietnam was made by TNK-BP in 2010. TNK-BP – the third largest oil company in Russia at the time that produced about 16 per cent of Russia's crude oil – was a joint venture between the British company BP and the Russian AAR Consortium (Alfa Group, Access Industries and Renova). Both BP and AAR owned equal shares (50 per cent and 50 per cent). TNK-BP also owned almost 50 per cent of another Russian oil and gas company – Slavneft ('Russia', 2010). Deputy Chairman of TNK-BP Maxim Barsky described the shipment to Vietnam as the 'first and important step towards broad-based cooperation between TNK-BP and PetroVietnam'. He continued that the company's goal was to establish a long-term presence in Vietnam and to pursue joint projects in the upstream and refining sectors ('Russia', 2010).

As a side note, TNK-BP was another example of how the presence of foreign oil companies was reduced on the Russian market: in 2013, BP's 50 per cent share in TNK-BP was acquired by Russia's Rosneft. After the deal, BP ended up owning 20 per cent of Rosneft's shares ('Companies & Markets; Rosneft Finishes Buying TNK-BP, Becomes World's Largest Publicly-traded Oil Co', 2013).

Siberian oil (the so-called ESPO blend) has been popular on the Asian markets and even in North America: as oil industry representatives described it:

> ESPO blend's good quality is attracting the attention of American refiners. While Alaskan crude currently accounts for 30 per cent of their loading, in the future Alaskan oil is likely to feel pressure from ESPO blend.
>
> (Besedovskiy, 2010)

The ESPO pipeline helped reduce the transportation distance by ocean tankers between Russia and the United States. Previously, the transportation distance for shipping oil from Northern Europe to the U.S. East Coast was roughly 40 per cent longer than the distance from Sakhalin to Alaskan ports and California. Because of this and other factors, the ESPO blend was cheaper than Alaskan, BP's, and Chevron's oil (Malik & Henshall, 2010). The Wall Street Journal reported in 2010 that ESPO oil helped keep California's gasoline prices low during the summer of 2010, providing an alternative to OPEC oil and increasing the energy supply of the U.S. West coast (Malik & Henshall, 2010).

The ESPO blend was also favored by refineries because of its 'sweet' quality: in industry terms, 34.7-degree API density and 0.53 per cent sulphur content. The blend was even considered superior to the popular Oman blend. In 2010, Tesoro Corp.'s CEO Bruce Smith expressed a strong interest in using the ESPO oil in the company's production of gasoline, diesel, and jet fuel (Malik & Henshall, 2010). Around that time, which was before the Western sanctions of 2014, the export of Russian oil to the U.S. West Coast soared from zero in 2009 to 100,000 barrels a day in 2010. Back then, oil analysts even predicted that ESPO's share could reach 15 per cent in the U.S. West Coast's oil imports (Besedovskiy, 2010).

Natural gas

Russia has the biggest reserves of natural gas in the world – 1,680 trillion cubic feet (Goldman, 2008). The country has even been called the 'Saudi Arabia of natural gas'. However, unlike oil trade, which is global, the feasibility of natural gas trade between countries depends on pipeline infrastructure and LNG technology. Because LNG is usually more expensive and the infrastructure of liquefaction has not been available everywhere, the natural gas trade has mainly depended on regional pipelines.

Only recently, LNG use has become more widespread, and it is shipped by tankers, similar to oil. An LNG supply chain connects remote natural gas fields with liquefaction plants on shore and marine loading terminals offshore. After LNG is transported by tankers, it is then re-gasified at regasification plants, providing natural gas in a gaseous form to gas distribution networks (power generating plants and industrial users).

Initially, with an eye on the U.S. domestic market, as well as Asian markets, Russia was focused on developing LNG facilities: for example, joint projects on the Sakhalin Island mentioned above. The first LNG plant was built on Sakhalin in 2009 as a part of the Sakhalin-2 joint venture. At its full capacity, the plant was producing five per cent of the world's LNG output at the time (Kramer, 2009). Another area in which Russia started to develop LNG has been the Shtokman natural gas field in the Arctic – a $20-billion project. Such expensive infrastructure inevitably makes LNG more expensive for foreign consumers than natural gas produced locally in their countries or delivered by regular pipelines. Currently, LNG is competitive only if consumers do not have enough natural gas domestically, or if other types of fuel (gasoline, ethanol, etc.) are more expensive than LNG. At the present moment, domestic natural gas production in the United States meets the domestic demand, because of shale gas technology, making LNG imports from Russia not economically feasible.

The reason the United States became self-sufficient in natural gas supply, as it was explained in Chapter 1, is because of the shale gas production that unlocked

'100 years' worth' of natural gas reserves. The Marcellus Shale and other shale gas sites are experiencing a new boom, similar to the oil boom a century ago. The daily production of shale gas in the United States has increased twenty-fold in just several years and reached 20 billion cubic feet of gas per day. The U.S. Department of Energy estimated in 2012 that the total amount of U.S. shale reserves is 482 trillion cubic feet ('United States: Shale Gas Boom to Boost Industry, Trade', 2013). Understanding this, Russia diverted its LNG exports to Japan, South Korea, Taiwan, Vietnam, and China.

In sum, Russia's enormous oil and gas resources until recently were exported mostly to Europe via an extensive system of pipelines and then by tankers around the world. To bypass transit countries and to ensure new outlets for the country's energy resources, Russia built new European terminals – e.g., the port of Primorsk, as well as a massive pipeline from East Siberia to the Pacific Ocean. The ESPO pipeline has also become a route that made Russian oil available to Asian consumers and the West Coast of the United States. This oil is highly competitive on the global market thanks to its low sulfur content and its moderate price, as well as low shipping costs via the Pacific Ocean.

In the natural gas trade, despite technological breakthroughs and multi-billion dollar facilities, Russia's LNG cannot compete with shale gas in the United States, which made the United States self-sufficient in natural gas production for many decades ahead. Russian projects in the harsh climate conditions of Sakhalin and the Arctic (e.g., the Shtokman field) potentially will pay off when LNG becomes a truly global commodity. Currently, LNG is in demand in well-to-do Asian countries (Japan, Korea, Taiwan, etc.), in which Russian gas competes with Australian and American LNG.

All in all, the economic potential of U.S.–Russia cooperation is much higher in oil trade and investment than in natural gas, and this potential remains unused, despite geographic and technological advantages. A worsening investment climate in Russia is one reason American companies have been leaving the Russian market.

The investment climate in Russia

After oil and gas reserves and transportation options, the next economic factor to consider in energy cooperation is the prevailing investment climate in a host country. The overall economic climate in a host country includes 'natural conditions and resources; infrastructure; human resources and expectations; and governmental and legal conditions' (Starr, 2003, p. 73). Investors in the energy sector also take into account political risks, especially if oil reserves are often located in countries with unstable regimes and a low level of democracy and market economy (Gati & Christiansen, 2003, p. 449).

Roughly, Russia's economic climate evolution can be broken down in several phases. The first phase, in the early 1990s, under President Yeltsin, was favorable for foreign investors who enjoyed large profits amidst Russia's chaotic economic domestic reforms. The second phase, in 1999–2002, came with the economic collapse of 1998, caused by an inefficient taxation system and the downfall of domestic production. The 'grey economy' was dominating the market, and most foreign investors downsized their presence in Russia. During the second phase, the early Putin administration briefly improved the investment climate, and the raging inflation of the 1990s was brought to manageable levels.

Phases 3 and 4 stand for the rest of Putin's presidency in 2002–2008 and Medvedev's presidency in 2008–2012. In 2010, Dmitry Medvedev realized that Russia was sliding down in the 'Doing Business' ratings. It was a period of the U.S.–Russia 'reset' and general warming in Russia's relations in the West. Medvedev often spoke about the need to diversify the Russian economy and to attract foreign investors. He introduced several initiatives with the goal of hiring more foreign workers who could bring expertise that Russia was lacking. Medvedev realized the need to improve the taxation system; to invest more in research and development; to modernize the customs system in Russia in order to make export-import operations more efficient; and to develop nano-technologies, which became his 'pet project'. He understood that without those reforms, Russia would become more and more isolated from the global economy, except for raw material exports. In just one year (from 2008 to 2009), the amount of foreign investments in the Russian economy decreased by 41 per cent, mostly because of the 2008 Russia–Georgia war (addressed in more detail below). The Russian government was trying to attract investors by reducing taxation for businesses and introducing the low 20 per cent tax for corporations ('Russian President Calls for Improving Investment Climate', 2010).

Finally, phase 5, started by Vladimir Putin's coming back to presidency for the third term, has been dominated by the ever-increasing state control over domestic resources and by the most serious geopolitical disagreements with the United States and the West since the end of the Cold War.

Because it is so important, the economic climate in Russia has been monitored and analyzed by foreign governments who provide recommendations and guidance to businesses. The U.S. government provides informational support for American companies to navigate in Russia's business environment. From the early 1990s, the *Doing Business in the New Independent States* guide encouraged companies to take advantage of the well-educated workforce in post-Soviet states and vast natural resources, as Russia was on its way to the market system:

> The emergence of a prosperous Russia as well as the other states will add billions of dollars of new growth to the global economy, which in turn will create exports, jobs, and investment opportunities in America.
>
> ('Doing Business in the New Independent States – A Resource Guide', 1994, p. 497)

Similar assessments were offered by other reports on Russia's economic conditions: the 'Doing Business' database and enterprise surveys by the World Bank's the *Business Environment and Enterprise Performance* survey and the *Russian*

Competitiveness and Investment Climate Assessment survey (Desai & Goldberg, 2008, p. 10).

In addition to informational support, the U.S. government-sponsored loans helped American companies start joint projects in Russia. Examples include a $2-billion U.S.–Russia oil and gas framework agreement and a $150-million U.S.–Russian joint venture, *Polar Lights*, sponsored by the Overseas Private Investment Corporation (the U.S. Government's development finance institution). In the 1990s, the U.S. Trade and Development Agency funded feasibility studies in the Russian oil and gas industry and provided training for Russian specialists ('Doing Business in the New Independent States – A Resource Guide', 1994).

Russia and the United States signed a number of bilateral treaties to reduce risk for investors, to guarantee profit repatriation, and to protect property rights. The first of them was based on the 1990 Trade Agreement with the Soviet Union and covered market access, tariffs, property rights, and other aspects (Novelli, 2003, p. 431). Shortly after that, the 1993 Oil and Gas Agreement provided the foundation for energy trade and investment, although the agreement was quickly followed by the Clinton administration's policy of *multiple pipelines* (with Richard Morningstar in charge of it), to reduce Russia's strategic influence as an energy exporter to Europe.

When those agreements were signed, the domestic economic situation in Russia was chaotic: energy deals were arbitrary and did not have a strong legal foundation, and investors were frustrated with corruption and the absence of clear laws, such as a clear PSA regime. All those negative factors were especially strong during Russia's economic crisis of 1998, when the amount of overall foreign direct investments (FDIs) in Russia fell sharply from $5 billion in 1997 to $3 billion in 1998. American FDIs fell from $3 billion to almost $2 billion over the same period. The economic difficulties of the late 1990s were one reason Vladimir Putin's becoming president in 2000 with a promise to double the country's GDP and to restore 'the rule of law', which was welcomed by the majority of the Russian population.

Vladimir Putin's first presidential term (2000–2004) coincided with an unprecedented growth of the Russian economy (4–6 per cent per year), a decreased level of inflation, and the external debt that was substantially reduced. Instead of the usual budget deficit of the 1990s, Russia had a $7-billion budget surplus already in 2001, through tax reforms and growing tax collection. Other reforms attempted by the Putin administration included a customs reform (aimed at simplifying and systematizing customs processes and to reducing red tape), a land reform (to strengthen property rights), and a currency control reform (for mandatory currency conversion requirements) (Marshall, 2003, pp. 13–19). Russia even became one of the world's three top-performing countries in stock value gains, earning an unofficial title of an economic 'tiger' in the post-Soviet region (Kalicki & Lawson, 2003, p. 473).

There is no consensus in the literature about the main reason of Russia's strong economic performance in the 2000s: two main explanations are Vladimir Putin's reforms and soaring oil prices that brought Russia unprecedented oil revenues. Russia's economic rise coincided with an increase in oil prices from $20 to $140 per barrel within that period. Marshall Goldman (2008), as well as Putin's economic adviser Andrei Illarionov (Maass, 2009), argues that oil income alone was enough for the country's progress, regardless of Vladimir Putin's reforms. Hillary Appel, on the contrary, argues that Vladimir Putin's personal leadership, when he reduced the influence of the key oligarchs and increased the state ownership of major industries, was the underlying condition for the domestic growth. Putin consolidated the state's fiscal power and removed Yeltsin-era tax evasion schemes, while restricting the influence of international institutions and foreign advisors (Appel, 2008, p. 302).

Most likely, both factors (high oil prices and Vladimir Putin's reforms) contributed to the growth in the Russian economy. Some of these reforms were introduced back in 2000, when oil prices were still low (about $18 per barrel). At the time, few analysts could predict that oil prices would soar to $140: '[o]ne could argue that oil prices zooming above $20 per barrel contributed to a false sense of economic security and, if taken to the extreme, acted as a disincentive to bolster the reform process' (Marshall, 2003, p. 17). This 'sense of security' skyrocketed when oil prices rose not to $20 but to $140 per barrel, bringing vast revenues to Russia and reducing incentives for the diversification of the economy.

Not surprisingly, since then, the Russian government has mainly focused on developing the oil and natural gas sector, including new oil fields, in order to increase oil production (Mankoff, 2009, p. 33), as well as diversifying oil and gas export routes, from Europe to the Pacific Rim (Hueper, 2003, p. 173). Even the goal to double the country's GDP, promised by Vladimir Putin in his election campaign in 2000, seemed possible to achieve when oil prices were so high (although this goal has never been achieved) (Mankoff, 2009, p. 32). The new wealth created a new elite, because oil 'often creates powerful incentives to capture and control the power and wealth that resource abundance provides', (Talbott & Montes-Negret, 2008, p. viii). The government was increasingly blending with the oil oligarchy. For example, Igor Sechin simultaneously held the positions of Rosneft's Chairman and vice prime minister of Russia. The oil sector was prioritized despite the fact that it employed only 1 per cent of Russia's workforce, compared to 15 per cent in manufacturing and 28.4 per cent in the service sector (Shaffer & Kuznetsov, 2008, p. 15). Inevitably, it led to increased income inequality, which was not protested by the population because the overall starndard of living increased.

Foreign investors also welcomed greater economic and political stability. In the fast growing Russian economy, they were able to receive much higher profits than in many other countries. In 2009, the U.S. Embassy in Moscow stated: 'Russia today is fundamentally a different country from what it was 10 years ago. The country's macro-economic indicators are stronger and healthier than at any time during the last 10 years' ('U.S. Commercial Services in Russia', 2009). Foreign companies were receiving high returns on their investments (600 per cent and higher), because the Russian economy was quickly expanding (Seib, 2008).

The positive estimation of investment conditions started to change as more contracts with foreign oil companies were revised in accordance with a new rule that a foreign share in joint energy enterprises must not exceed 50 per cent, and Russian partners must have a controlling stake. Contracts from the 1990s with a foreign share of more than 50 per cent were revised, and the excessive ownership rights were transferred to Russian companies. For example, in 2007, Shell sold a share in Sakhalin-2 to Gazprom for $7.45 billion (Simonov, 2011). Those revisions brought additional risks for foreign investors. In 2010, CEO of Chevron Neftegaz (a Russian branch of Chevron), Darrell Cordry said that this 'strategic assessment appraisal process should be done up front, before the exploration period commences, and national security issues should be considered at the time' ('Chevron Concern about Russian Law on Foreign Investment and its Impact on Oil Exploration', 2010).

Trade between Russia and the United States remained asymmetrical. The United States was the third biggest importer for Russia (8 per cent of Russian import came from the United States) and the seventh biggest destination for Russian export (4 per cent of Russian export went to the United States). Russia, however, was only the 30th on the list of U.S. largest trading partners. In addition to oil, top exports to the United States from Russia mostly included commodities: palladium, enriched uranium, oil, diamonds, and vodka. Top American exports to Russia included tobacco, high-tech equipment, automobiles, and other high-tech goods (Lotarski, 2003, p. 30). These rankings and structure have not changed since then, although they were affected by the Western sanctions of 2014–2015.

Russia's oil revenues have not been conducive for developing other sectors of the economy and reducing Russia's overwhelming dependence on oil and natural gas. One attempt to diversify the economy was made by President Dmitry Medvedev in 2008–2012. One of his major projects – Skolkovo – was designed to create a modernized high-tech enclave, similar to the Silicon Valley in California. Medvedev was hoping to give a boost to industries outside the energy sector, promoting innovative development, with special attention to nanotechnology. Billions of dollars were invested in Skolkovo, with such agencies as the Modernization Commission, the Skolkovo Council, and the Skolkovo Foundation in charge. Medvedev's aspirations to revitalize the Russian economy were accompanied by his affinity for modern technology and social media. Still, experts have expressed doubt that the top-down approach can instill the Silicon Valley entrepreneurial culture into Russia's technology domain. Some have argued that forcing the start-up culture is destined to fail. The Skolkovo project has been led by billionaire Viktor Vekselberg and his family, which again was an example of the oligarchic nature of the project (Wadhwa, 2010).

The future of the economic diversification trend remains unclear, as Russia and the West imposed mutual economic sanctions after the events in Ukraine in 2014–2015 (explored in more detail in Chapter 5). Before the sanctions, Russia's dependence on Western goods and services was favorable for the balance of power with the West, because an energy supplier 'may become dependent on the goods it is importing with its profits' (Quester, 2007, p. 446). Therefore, experts noted that Russia's dependence on Western goods provided a guarantee against a comeback of communism, because people will not want to lose their standard of living (Quester, 2007, p. 448). Also, Western oil companies had a large market in Russia for their drilling technology and for oil and gas exploration.

International energy companies in Russia

As it was mentioned above, international energy companies (Chevron, ExxonMobil, ConocoPhillips, and others) started to work in Russia in the 1990s, signing PSAs with the Russian government (Darmin, 2010, p. 31). However, since Vladimir Putin initiated his reforms in the 2000s, PSAs became less common. Foreign companies have participated mostly in technologically and financially challenging oil and gas projects, such as those in the Arctic. For more than two decades, the companies have been navigating in Russia's changing business environment.

Chevron

Chevron has participated in oil production and transportation in the Caspian region, the Sakhalin Island, the Shtokman field, and Yamal. In 1996, Chevron signed its first major contract in Russia, acquiring one third of the Caspian Pipeline Consortium (CPC) – a 935-mile pipeline that connects the Tengiz oil field in Kazakhstan and the Russian port of Novorossiysk ('Chevron Neftegaz. Inc', 2000). Overall, Chevron's invested $2.2 billion in CPC, acquiring 15 per cent of the consortium by 2006. In 2001, the Consortium's CPS oil pipeline started to transport 700,000 barrels of oil per day. Chevron continued to increase its participation in the Consortium, investing $710 million in Tengizchevroil and bringing its total investments in CPC to $4.3 billion in 2010 (Chevron, 2010).

In the 1990s, Chevron joined Gazprom Neft in the Severnaya Taiga Neftegaz LNG project in the Arctic. However, Chevron withdrew from the project because the output of natural gas was lower than it had been predicted ('Chevron i Gazprom Neft Obsuzhdayut Proekti', 2009). The company's CEO, Darrell Cordry has reiterated on several occasions that Chevron is still interested in other LNG production projects in the Arctic jointly with Gazprom and Novatek, such as the Yamal LNG on the South Tambeyskoye field ('Chevron, Gazprom Identify Possible Joint Projects', 2009).

In addition to crude oil transportation, Chevron has operated on the Russian market of oil lubricants (their production, licensing, and technology transfer), cooperating with LUKoil, OAO Gazprom Neft, TNK-BP, and other Russian lubricating oil producers. Chevron licensed its lubricant, diesel, and jet fuel production technologies to a number of Russian companies, including Kirishi Nefteorgsyntez (Surgutneftegaz), LUKoil-Volgograd-Neftepererabotka, Taftneft, Rosneft, and several refineries in Tuapse, Komsomolsk, Achinsk, Novokuibyshevk, and Ryazan (Chevron, 2010).

During Vladimir Putin's first two terms in 2000–2008, because of the imposed restrictions on foreign participation in the Russian energy sector, Chevron was not able to start new major large-scale projects. Only during the 2010 'reset' in U.S.–Russia relations did Chevron manage to sign a major deal with Rosneft. The two companies agreed to jointly develop an offshore 860-million-ton oil field in the Black Sea. Initially, they invested $1 billion each and further committed to invest up to $32 billion for the project implementation ('Chevron Signs Project in Russia', 2010).

As an additional condition for such large projects, the Russian law requires that the oil company invest in local infrastructure: medical, educational, and recreational facilities for the local communities. To further boost its public image in Russia, Chevron spent about $4 million from 1994 to 2010 on cultural projects, sponsoring the Russian National Orchestra, the State Tretyakov Gallery, the State Pushkin Museum, the State Literary Museum, the Moscow Zoo, and the Moscow Kremlin museums. Chevron also maintains long-term partnerships with Russian universities such as the Moscow State University, VNIIUS (Kazan), Gubkin State Oil and Gas University, the All-Union Geological Oil Exploration Institute (VNIGNI), and the Scientific & Research Institute of Natural Gases and Gas Technologies (VNIIGAZ), providing financial help to students and sponsoring conferences (Chevron, 2010). Particularly thanks to such initiatives, the company has maintained a stable presence in Russia.

ExxonMobil

ExxonMobil followed a path in Russia similar to Chevron's. In 1995, ExxonMobil signed a PSA for oil and natural gas production in the Sakhalin Island, investing $4.7 billion over several years. This project, known as Sakhalin-1, started as a Russian–American–Japanese joint venture. Currently, it is co-owned by Exxon Neftegas Limited (a Russian branch of ExxonMobil) – 30 per cent of shares, USA; the Sakhalin Oil and Gas Development Co. Ltd. (30 per cent, Japan), Rosneft RN-Astra (8.5 per cent, Russia), Sakhalinmorneftegas-Shelf (11.5 per cent, Russia), and ONGC Videsh Ltd. (20 per cent, India) (Taylor, 2010). Sakhalin-1 develops the Chayvo, Odoptu, and Arkutun-Dagi oil fields, which possess 2.3 billion barrels of oil and 17.1 trillion cubic feet of natural gas in reserves.

These fields were first discovered in the 1970s, but they were undeveloped for several decades because Russia was lacking technology and investment ('Russia-Sakhalin-1', 2010). Sakhalin-1 was the first step in ExxonMobil's planned $12-billion investment in the Russian energy sector (Marshall, 2003, p. 15). Production of oil and gas in Sakhalin has required ExxonMobil to use its best technologies, in order to overcome harsh climate conditions (pack ice, waves, earthquakes) and seismic activity ('Russia-Sakhalin-1', 2010).

Oil production in one of the projects above – the Chayvo field – started in 2005, followed by a new onshore processing facility in 2006. By 2010, this field alone produced 210 billion cubic feet of natural gas and 270 million barrels of oil: 250,000 barrels per day, shipped by a fleet of 350 tankers from the De-Kastri terminal (Khavarovsky Krai, Russia). The project passed the cost recovery point in 2008 and brought in about $4 billion of royalties and taxes to the Russian state budget. The next step by ExxonMobil is to expand oil production into the Odoptu and Arkutun-Dagi fields and to start producing natural gas (Taylor, 2010).

Like Chevron, ExxonMobil has complied with Russia's labor safety regulations, environmental protection laws, and it responded to the indigenous peoples' concerns. One of the requirements by the Russian government was that 75 per cent of the company's total 13,000 staff in Russia should be Russian specialists. Since Sakhalin-1 is located in an underdeveloped region, the company also contributed to new local infrastructure: roads, bridges, airports, sea ports, and education, medical, and community facilities. Assisted by the United States Agency for International Development (USAID), ExxonMobil provided training and microloans (about $1 million in total) to local entrepreneurs who open their own small businesses in the area (Taylor, 2010). President of Exxon Neftegaz, James Taylor underscored in 2010 that Sakhalin-1 provided 'jobs for locals and new contracting opportunities', as well as revenues for the Russian budget. He also stated that Sakhalin-1 helped expand the natural gas availability for the consumers in the Russian Far East, most of whom up to that point had had to rely on expensive electricity for heating and cooking (Taylor, 2010).

Similar to other international companies, however, ExxonMobil was not allowed to access new contracts in the Russian oil and gas sector for almost a decade during Vladimir Putin's first and second presidential terms. Only in 2011, a major breakthrough happened for the company, when it signed a multi-billion-dollar deal with Rosneft to produce offshore oil in the Kara Sea and in the Black Sea. In response, the deal also gave Rosneft limited access to oil fields in the United States in the Gulf of Mexico ('Rosneft Teams Up with ExxonMobil in Arctic Deal', 2011).

ConocoPhillips

ConocoPhillips came to Russia in the 1990s and invested about $11 billion by 2010 in such joint projects as Polar Lights Company (with Rosneft in Timan Pechora). In 1995, the company partnered with Total (France), Hydro (Norway), and Fortum (Finland) in an international consortium established for the Shtokman gas field development. In 2002, however, their license for this project expired and the contract was instead awarded to Rosneft and Gazprom (Stern, 2005, p. 17).

In 2005, ConocoPhillips established the Naryanmarneftegaz (NMNG) – a limited liability company between LUKoil (70 per cent) and ConocoPhillips (30 per cent). Similar to Chevron and ExxonMobil, ConocoPhillips has always emphasized its strong adherence to Russia's environmental and safety regulations in oil and natural gas production processes, 'aimed at zero-incident operation of hazardous oil and gas production facilities in the extreme climatic conditions of the North' ('Projects in Russia', 2010).

ConocoPhillips and LUKoil have closely cooperated in multiple ways, and ConocoPhillips even bought 20 per cent of LUKoil's shares. In addition to domestic production in Russia, the two companies jointly participated in international bids, such as West Qurna-1 in Iraq. They lost the bid, which was awarded by the Iraqi government to ExxonMobil and Royal Dutch Shell. ConocoPhillips and LUKoil were supposed to bid together for West Qurna-2, but at that time (2009) LUKoil chose Statoil as a partner, because of the difficulties that American companies were experiencing in Russia (Tutushkin & Vasilyev, 2009).

In 2010, ConocoPhillips sold 10 per cent of LUKoil shares, worth $4.7 billion, explaining the move by the unfavorable investment climate in Russia, in which state-owned companies (Rosneft, Gazprom) receive the most lucrative contracts and oil fields. The company had its taxes in Russia increased without notice, and its license was revoked from time to time in the Polar Lights joint venture. After taxes, fees, and informal 'payments' to local officials, the company's profits were lower than expected (Goldman, 2008, pp. 84–85). ConocoPhillips was also struggling during the global recession, in which the company lost $17 billion, and in addition to the Russian LUKoil assets, the company sold shares in a Canadian energy business and a pipeline in the United States (Ryabova, 2010).

Geopolitical rivalries

The business opportunities for American companies mentioned above have been affected not only by the investment climate but also geopolitical rivalries that led to the Russian government's tightening of the energy markets. The last decade was a period of difficulties and disagreements between the Russian and American governments. Such issues as NATO's expansion, the war in Iraq, several discontinued international treaties (such as the Anti-Ballistic Missile Treaty (ABM), the U.S. missile defense plans in Europe, the Iranian nuclear issue, and the Russia–Georgia War of 2008 were affecting economic relations between Russia and the West. Such events affected business opportunities in the strategically vital energy sector in Russia. Especially during the 2008 Russia–Georgia War, the relations between the Putin administration and the Bush administration were so tense that analysts almost considered it as the beginning of a new Cold War.

The Cold War legacy

Experts agree that not only the Cold War itself, but also interpretations of how it ended matter for contemporary U.S.–Russia relations. In Russia, the prevailing view is that the Cold War ended as a 'negotiated outcome that benefited both sides'; in the United States it is commonly viewed as a *defeat* of the Soviet Union and a 'necessary struggle ... that culminated in a decisive, final victory of American moral and political principles...' (Foglesong, 2007, p. 4).

Along these lines, experts on Russia who spent extended periods of time in the Soviet Union when it was disintegrating (such as Robert English) provide convincing evidence that the Soviet Union ended as a result of new ideas originating from *within*, rather than simply due to military and economic pressures by the United States (English, 2000). Another scholar and practitioner – Jack F. Matlock, who was the U.S. Ambassador to the Union of Soviet Socialist Republics (USSR) in 1987–1991 agrees with Robert English: 'pressure from government outside the Soviet Union, whether from America or Europe or anywhere else, had nothing to do with it', as he refers to the breakup of the Soviet Union (Matlock, 2010, p. 89). In Matlock's view, the end of the Cold War resulted from extensive negotiations and mutual compromises, in which he participated first-hand. In those negotiations, the last president of the USSR Mikhail Gorbachev was willing to reform the country, to unify Germany, and to cooperate with the United States. In return, the U.S. government promised not to expand NATO to the East. Matlock emphasizes that referring to Russia as a *defeated* enemy in the Cold War is detrimental to the U.S.–Russia cooperation.

The Cold War legacy still affects political rhetoric on both sides and leads to unnecessary confrontation. For example, Russia's market reforms in the 1990s have been named by some analysts as an American 'crusade' and a modernization project (Cohen, 2001). The post-Cold War rhetoric in the United States has focused on the 'Russian autocratic tradition' that needed to be corrected. The Cold War references were used in order to justify 'new' European countries' joining NATO, in search of protection against Russia. Both in external affairs, such as NATO, and international market reforms guided by American advisors, the predominant Russian view was that the 'shock therapy' reforms of the 1990s were a crime on humanity, in that they pushed the majority of Russians into poverty and despair (Lieven, 2000).

NATO *expansion*

As the American involvement in advising Russia's 'shock therapy' in the 1990s became a distant history for most Russians, another issue that has never gone away as an irritating factor for Russia has been the expansion of NATO to the East, directly to the Russian borders. NATO's presence in its current borders and a possible inclusion of Ukraine and Georgia, which was seriously discussed in the United States in 2008, have been fueling the anti-American sentiments in Russia. NATO's expansion was a salient issue to 'rally around the flag' (Bazhanov, 2010). In 2006, Elena Khalevinskaia of the Russian Institute of World Economy summarized the Russian view on NATO's expansion: the United States has used it as a method to 'belittle the role of Russia as one of the influential centers of the budding multi-polar world' (Piadyshev, 2006, pp. 21–25). In the same year, another Russian analyst, Vladimir Podoprigora, argued that NATO's expansion has been an arms race and a method aimed at tying NATO members to American weapon standards and creating their dependency on U.S. military supplies and maintenance (Piadyshev, 2006, pp. 21–25).

Not everyone in Russia, however, completely rejects the positive sides of NATO's expansion. Yevgeny Bazhanov, Vice Chancellor of Research and International Relations at the Russian Foreign Ministry's Diplomatic Academy wrote in 2010 that Russia could even benefit from a strong NATO, as the alliance could bring peace and stability to Europe. He continued that even those in Russia who called NATO the 'enemy' were still 'the same people who gladly take vacations in NATO countries, buy real estate there, send their children there to study and even root for those countries' football teams' (Bazhanov, 2010). He concluded that NATO, as a European common security mechanism, reduced rivalries and built trust among European countries, which was favorable for Russia, as well.

NATO's expansion has been closely connected to another contentious issue – the U.S. missile defense system in Eastern Europe. A decade ago, the proponents of the plan to place the system in the Czech Republic were very vocal in the United States, and they were perceived extremely negatively in Russia. Russia claimed that the system was designed to radar the Russian territory, rather than protect Europe against 'rouge states'. In response to the U.S. missile system plans, Russia announced plans to increase its own missile capabilities and possibly an antimissile system of their own (Piadyshev, 2006, p. 19). Some analysts saw the U.S. missile defense system as a negotiation tool for Russia: they argued that Russia perfectly understood that the system was not a threat: 'Russia privately recognizes that missile defense as planned does not currently pose a challenge mainly because the location, range and number of planned interceptors are too far, too short and too few to undermine Russia's strategic nuclear forces' (Sueldo, 2011).

Even though President Obama has not pushed the issue of missile defense in Europe lately, it is still an irritating issue for Russia, which could potentially lead to a new round of arms race between the United States and Russia. As recently as in May 2015, at the United Nations nonproliferation conference, an attempt to 'expand the world's premier arms-control treaty ended in failure' partly because of the mistrust between Russia and the United States. The discussion of the Treaty on the Non-Proliferation of Nuclear Weapons, which has 189 participating states, focused on the Russia–Ukraine conflict among other issues, and the Russian representative Mikhail Ulyanov referred to the U.S. missile-defense programs in Europe as 'a serious obstacle to further disarmament' (Zak, 2015).

After the Georgia war of 2008, and now the Ukraine conflict, the idea of any reductions in defense spending is not popular in Russia. The conversion of military industries for civilian use, which has cost Russia over $150 billion since the 1990s, is considered too costly and not beneficial for Russia's interests (MacLean, 2006, p. 29). Russia's Tactical Missiles Corporation, for example, increased its sales 118 per cent just in 2013. Other arms trading companies followed suit: United Aircraft Corporation increased its volume of sales by 20 per cent, and Almaz-Antey – by 34 per cent. Arms and military expenditure expert Siemon Wezeman described these increases as:

> The remarkable increases in Russian companies' arms sales in both 2012 and 2013 are in large part due to uninterrupted investments in military procurement by the Russian Government during the 2000s. These investments are explicitly intended to modernize national production capabilities and weapons in order to bring them on par with major US and Western European arms producers' capabilities and technologies.
>
> ('Russia Armaments', 2015)

In addition to conventional and nuclear weapons, Russia is currently investing in creating drones and combat robots, as well as in cybersecurity ('Russia Armaments', 2015).

Demilitarization has been especially unpopular in Russian regions heavily dependent on military industries. Governor of one of them, the Saratov oblast, stated back in 2006 that the geopolitical situation had not changed much to the better since the Cold War. Echoing a common sentiment in the Russian military industry, he maintained that military research and development would encourage Russia's technological progress in other sectors of the economy (Piadyshev, 2006, p. 22).

Rivalry in Eurasia and Russia's energy policy in the Former Soviet Union

U.S.–Russia rivalries have been especially strong in Russia's 'backyard': the Russia's 'Near Abroad' and the Middle East, as well as in Europe (Starr, 2003). In these geographical areas, Russia has a strong sense of geopolitical urgency and insecurity, often challenged by other great powers.

This rivalry has been especially noticeable in the area of energy security. In the 1990s, to diversify energy supplies for the West and to ease Europe's dependency on Russia, the United States started to promote the policy of multiple pipelines. These developments culminated in the 1990s in the Contract of the Century in Azerbaijan, to build the Baku-Tbilisi-Ceyhan (BTC) pipeline from Azerbaijan to Turkey via Georgia (more on this in Chapter 3). BTC was a blow to Russia's interests, snatching away Russia's monopoly for transportation routes in the Former Soviet Union. This pipeline, as well as the Caspian Pipeline Consortium from Kazakhstan built in 2001, was the great powers' methods to push for their 'Great Game' for the influence in Central Asia and the Caucasus (Starr, 2003). In addition to trying to control as many transport routes for oil and gas as possible, Russia has continued to supply the former Soviet republics – Armenia, Belarus, Ukraine – with subsidized energy resources, in order to exert Russia's influence.

The U.S. policy of multiple pipelines continued under the Bush administration as part of the National Energy Policy, announced in 2001 by Vice President Dick Cheney. The goals of the policy were to establish relations with various producers of energy, to diversify energy sources for the United States, and to reduce the influence of Russia and OPEC (Mortished, 2008). South Caucasus became a strategically important region for such a diversification, in an attempt to break the Russian transport monopoly (Mann, 2003).

Despite the policy of multiple pipelines, Russia closely cooperated with the United States after the terrorist attack of September 11, 2001. Russia and the Central Asian states supported the U.S. military operation in Afghanistan, providing intelligence and the air space access. Over time, however, the presence of American military bases in Russia's backyard made the Russian government worry about the long-term strategic and economic implications and a possible shift in the balance of power in this important region, affecting energy resources and transit routes (Gati & Christiansen, 2003, p. 453). Simultaneously, Russia had to 'swallow' the U.S. withdrawal from the ABM Treaty in 2002. The disagreements became an especially salient issue when Russia opposed the war in Iraq in 2003. At that point American policymakers said that Russia was not providing enough support for the U.S. goals: to fight the war on terror, to promote democracy, and to stop nuclear proliferation.

Russia's disapproval (together with France and Germany) of the 2003 Iraq War was a large step in a downhill slide in U.S.–Russia relations. By 2006, cooperation became the 'exception, not the norm' (Mankoff, 2009, p. 6). In the Western political rhetoric, Russia was often described as a rival to the United States and the European Union, trying 'to subvert its neighbors and prevent the expansion of free-market democracy around its borders' (Mankoff, 2009, p. 6). On the other hand, the Russian government accused the United States of unilateralism (Goldgeier & McFaul, 2005, p. 45). Russian experts criticized the United States as a 'world overlord', praising Russia's desire to resist a one-polar world (Piadyshev, 2006, p. 31).

Vladimir Putin set a goal to restore Russia's great power status after the domestic chaos of the 1990s. The first decade after the dissolution of the Soviet Union was, as Deputy Foreign Minister Andrei Denisov put it, when 'under various pretexts, democratization being one of them, our international partners and rivals' were 'pursuing their own obvious and selfish aims' (Piadyshev, 2006, pp. 18–19). A major theme in Russian rhetoric was '[s]trong states are loved, strong states are respected' (Piadyshev, 2006, p. 19).

Russia's economic recovery in the 2000s was 'shocking mostly for its rapidity' to many in the West. The West came to see a newly 'resurgent' and assertive Russia, which wanted to be a great power again and whose 'integration with the West and its institutions was neither possible nor desirable' (Mankoff, 2009, p. 4). Energy resources – the main source of Russia's wealth and influence – were now a source of conflict, as Aleksandr Lebedev of the Russian State Duma stated in 2006:

> [t]ensions will grow together with the world's population while the energy resources, an object of uncompromising rivalry, will decrease. Russia that controls a large part of the world's natural resources will, therefore, be challenged and should be prepared to use its military shield to survive [sic].
>
> (Piadyshev, 2006)

Access to energy resources around the world has become an even more contentious issue between Russia and the United States. Russian experts, such as Vagif Guseinov, called it 'resource imperialism' and argued that the United States aimed at controlling world resources through the unipolar system and 'color' revolutions. He continued that the world's 'democratic revolution', promoted by the United States, has been in fact designed to 'cement' the new imperialism, backed by the technological, military, and economic superiority of the United States (Guseinov, 2008, pp. 438–440). In the West, on the contrary, Russia has been viewed as an energy supplier that has repeatedly exploited its 'energy weapon', creating dangerous dependencies on Russian oil and gas in Europe.

Russia's military was 'tested' in 2008 during a war with Georgia – a small transit state that had a strategic position in the regional energy security: 'Georgia was a stunning announcement that Russia had again become a force to be reckoned with' (Mankoff, 2009, p. 1). For Russia, the conflict was strong evidence that the country needed to build up the conventional weapons, as the nuclear arsenal alone was not able to deter smaller military conflicts such as the Russia–Georgia War: '…these weapons will not be much use in dealing with problems such as the situation in the North Caucasus' (Friedman, 2011). During and immediately after the war, most American and East European politicians, such as U.S. Senator John McCain, called for cutting off all relations with Russia and for excluding the country from international organizations. Others, such as Jack Matlock, insisted that negotiations should be continued (Matlock, 2010).

Another expert, who worked in the Soviet Union during the Reagan administration, Suzanne Massie, agreed with Matlock's view. As an advisor to President Ronald Reagan, she recalls how the U.S.–USSR relations were transformed during 1983–1988, from the lowest point and the 'evil empire' rhetoric to Reagan's 'walking around the Red Square arm in arm with Mikhail Gorbachev' (Massie, 2008). Analyzing this turnaround in Reagan's thinking, Massie recalls how Reagan distinguished between the Soviet government and the Russian people; he learnt about what the Russian people were thinking and experiencing, as human beings, not just a 'monolith marching as one toward a glorious communist future' (Massie, 2008). Based on those historical lessons, both Massie and Matlock insisted that negotiations with Russia be continued, instead of full isolationism.

Even though the Russia–Georgia war did not damage any pipelines in Georgia, Russia's reputation as an energy supplier had been tainted during natural gas transit disputes with Ukraine and Belarus: 'Demanding market prices from Kyiv and other post-Soviet capitals allowed Gazprom to realize higher revenues and to sow political chaos in countries seen as turning their backs on Russia' (Mankoff, 2009, p. 37). European experts mentioned that both sides (Russia and Ukraine) were to blame for the interruption of gas transit to Western Europe:

> Rather than Russia exploiting its uniqueness as the original source of this energy, Belarus and Ukraine may have been conversely exploiting their unique geographic position as transit countries (as well as exploiting the

> simple inertia of paying lower prices since the Soviet Union collapsed and the world's sympathy for anyone suddenly cut off from such a subsidy), with Western receivers being inconvenienced by this contest of wills between Moscow and Kyiv [sic].
>
> (Quester, 2007, p. 448)

The Western commentators called the disputes as Russia's energy blackmail of the European Union and the West and Russia's method to recreate the Soviet style of domination. Others, however, maintained that Russia was acting as a 'normal' great power, using carrots and sticks in negotiations post-Soviet states, with no designs to question their sovereignty or to establish 'an empire' (Tsygankov, 2005). The claims to sovereignty were being tested during Russia's annexation of Crimea, when a referendum in Crimea resulted in the majority of voters supporting the decision (these events are discussed in detail in Chapter 5).

Military-economic relations with third countries

Russia's relations with third countries have had serious repercussions for U.S.–Russia relations, as well, since Russia's and U.S. policies toward such countries as Iran, Syria, Libya, and others have often been opposite to each other.

In addition to the concerns about Russia's domination in the 'Near Abroad' and Europe, the United States has also been concerned with Russia's energy policies in the Middle East, especially in Iran and Iraq. Unlike the United States, Russia has been supportive of the Iranian nuclear program for years, and Russia's divergence from the U.S. position on nuclear cooperation with Iran has been another obstacle in U.S.–Russia cooperation. Compared to Russia's policies in Iran, Russia has followed the U.S. leadership in Iraq. In 2009, Russia even agreed to forgive the $12 billion Iraqi debt in order to try to secure the West Qurna project. Amidst concerns about the access to Iraq's oil and gas, Russian senator Margelov praised the fact that the United States allowed thirty-five other countries to participate in tenders for oil and gas development (Anishyuk, 2009).

The Iranian nuclear problem, mentioned above, has been a major stumbling block between the United States and Russia in the Middle East. Although the United States and Russia have repeatedly negotiated on the Iranian issue, it has remained unsolved, and Iran has continued its uranium enrichment activities. Russia's long-time cooperation with Iran and resistance to Western initiatives has been puzzling, considering Russia's desire to be an equal partner in the G-8 club of states and Iran's defiance of the international community. Nevertheless, for two decades, Russia has participated in the Iranian nuclear program, opposing Western sanctions imposed on Iran. Russia's support of Iran seems to contradict Russia's other goals: the Russian government has repeatedly stated that it was *not* interested in another nuclear state in the Middle East, Russia's 'backyard'. Nor is Russia-Iran cooperation simply revenue-driven, because other sources (oil and natural gas exports) in Russia's state budget are much more substantial. Most importantly, the cooperation with Iran alienates the United States and Europe (Parker, 2012).

Another issue that has been very problematic in U.S.–Russia relations is the Russia–Syria arms trade. The Syrian regime has cracked down on its own citizens in 2011, killing thousands of people. Despite this brutality, Russia has refused to support the regime change in Syria and to stop arms sales to this country. Russia's position on this issue has been the opposite to that of the United States and other Western countries. Instead, Russia's position was in 'unyielding alignment' with Syria and that they 'provided a diplomatic shield for Damascus in the UN Security Council' (Allison, 2013).

Syria has not been the only case in which the United States and Russia have disagreed on arms trade and technology transfer and arms (Trenin, 2002, p. 204). In a similar way, Russia has been supplying China with modern military technologies, even though China and Russia are potential geopolitical rivals. International relations scholars explain Russia's military cooperation with China by Russia's declining relative structural position, domestic politics factors, and the pursuit of short-term revenues (Donaldson & Donaldson, 2003). More recently, China has also been increasingly interested in buying oil and natural gas from Russia, substantially reducing Russia's dependency on Western consumers (Mortished, 2008). Japan is also a consumer of Russian oil, despite the fact that Russia and Japan have had a half-century dispute over the Kuril Islands and the 196,000-square km economic zone around them, which has prevented Russia and Japan from signing a peace treaty after WWII. The Kuril Islands became an issue of national pride for the Russians, and it is unlikely that Russia would make concessions to Japan (Trenin, 2002, p. 218).

The United States has also been concerned about Russia's deals with other NATO members. As Russia has balanced between its integration into the world economy and its 'ability to project hard power' (Mankoff, 2009, p. 2), the U.S. government has often disagreed with how NATO members deal with Russia, especially in the military sector. In 2010, the United States strongly objected when France sold to Russia Mistral-class warships, able to carry helicopters. American analysts indicated that the ships would give Russia 'additional capabilities to threaten Georgia' (Socor, 2010). Analyst Vladimir Socor stressed that such deals undermine NATO, 'enhancing Russia's capacity to pressure NATO allies and partners in Europe's East' (Socor, 2010). In 2015, however, France have decided not to give Russia the built and paid warship, because of the Western sanctions after the events in Ukraine ('Russian foreign minister interviewed on EU sanctions, Ukraine, Kuril Islands', 2015).

All the events above were influenced by domestic interest groups that were lobbying and shaping the governments' decisions.

Domestic interest groups

Energy trade and investment are more likely to happen successfully if powerful domestic groups and lobbies benefit from them. However, if interest groups are divided, the state's autonomy in pursuit of energy security via energy policy formations will increase. In the case of U.S.–Russia, the following groups have exercised their influence on the government policies: Russian political elite groups, Russian energy lobbies, and American business lobbies in Russia. Even though there is a certain degree of overlap among these groups, the insight into their specific influence helps explain the turbulence of U.S.–Russia relations.

Russian political elite

The Russian elite circles can be categorized along several dimensions according to their attitude to the West. Foreign policy in Russia is greatly dependent on the type of political elite in charge.

The first intellectual movement – *integrationists* – views Russia as a European state and a part of the West. They would like to adopt Western values and ideas, such as democracy and grassroots participation. These so-called 'Westernizers' were especially active and powerful in the early 1990s, when Russian politics had an almost unconditional pro-Western orientation (Melville & Shakleina, 2005, p. x). Among such Westernizers were the last president of the USSR Mikhail Gorbachev, Foreign Minister Andrei Kozyrev, and Deputy Foreign Minister Anatoliy Adamishin. This set of ideas was essentially adopted by the first president of Russia, Boris Yeltsin, who 'dampened Russian revanchism, jingoism, and nostalgia for the Soviet Union' (Colton, 2008, p. 266). In the last decade, influential voices in this intellectual branch include scholars Andrey Piontkovsky and Pavel Felgenhauer. Currently, Westernizers influence a small circle of public elite and have little influence on Vladimir Putin's policymaking.

The second category can be described as *balancers* – they emphasize Russia's role as a great power and a 'geopolitically and culturally distinct entity' that needs to balance the United States in the world, while pursuing limited cooperation. In the 1990s, Foreign Minister Yevgeniy Primakov was a typical representative of this category (Tsygankov, 2005, p. 136).

The third category was described by the international relations literature as *neo-imperialists*, such as the Communist Party leader Gennady Zyuganov. Since the beginning of the 1990s, they argue for the restoration of the lost power and for the independence from the 'alien' West (Tsygankov, 2005, p. 137). Currently, the political parties of this school of thought – the Communist Party of the Russian Federation (KPRF), the Liberal Democratic Party of Russia (LDPR) (led by Vladimir Zhirinovsky), and 'A Just Russia' party (led by Oleg Shein) – are in the so-called 'left-wing opposition' to the Putin administration, and the analysts have repeatedly noted their low profile. Although their ideas about 'equitable distribution of oil revenues, the preservation of the welfare state, and control over official salaries' are favored by the Russian public, this movement does not have enough political power and in fact is not even a challenge to Vladimir Putin. Russian analyst Mikhail Magid succinctly explained why this school of thought is in line with the official politics of the Russian government:

> Nostalgia for the Soviet Union – that dictatorship and mighty empire which challenged the West and forced its own peoples (Chechens, Lithuanians, Ukrainians, Hungarians, Poles, Czechs) into submission – that's what motivates most Russian voters and, indeed, most people on the left. And from that perspective, everything the Kremlin has done and continues to do is correct.
>
> ('The Kremlin has Nothing to Fear from Left-wing Opposition', 2014)

Finally, another branch, the *great-power normalizers*, have become the mainstream political sentiment in Russia. They are driven by moderately nationalistic ideas, a selective engagement in the post-Soviet space, with a strong emphasis on Russia's great power status. Such scholars as Andranik Migranyan and Alexei Arbatov can be categorized as *normalizers*, but most importantly, this philosophy became the basis of Vladimir Putin's presidency (Tsygankov, 2005, p. 138). The Russian government has been working on restoring the great power status in both foreign policy and domestic issues. One of the latest domestic laws has been a bill to ban 'undesirable' foreign and international non-governmental organizations (NGOs) from operating in Russia. The NGOs may be deemed 'undesirable' on the grounds of national security considerations and the 'foundations of Russia's constitutional order, defensive capacity and security' ('Russia's Putin Signs Law on "Undesirable" Foreign NGOs', 2015).

These four groups provide an intellectual basis to debates in Russia about possible paths. When it comes to the actual decisions about specific energy deals, a more practical group comes into play – representatives of Russian energy companies, who are often government officials themselves.

Russian energy lobby

As such, the energy policy in Russia is decided by a small elite circle (Mankoff, 2009, p. 9). The energy business is tightly intertwined with the government: government members are also board members of major state-run energy companies: 'crude oil and political power are umbilically connected in Russia' (Maass, 2009, p. 190). For example, Valery Yazev is a policymaker who has held leadership positions in energy businesses, the Russian legislature, and professional organizations (e.g., the Russian Gas Society) (Yazev, 2010).

Because of such close ties between the industry and the government, scholars define the Russian political system as oligarchic capitalism: a 'system in which prominent businessmen command political power, and powerful politicians are able to arbitrarily choose winners and losers' (Hoffman, 2003, p. 372). As Vladimir Putin consolidated the power of the state, he made sure that oligarchs did not exceed their power, as it was defined in the informal social contract. Several major oligarchs (e.g. Vladimir Gusinsky and Boris Berezovsky) who did not agree with Vladimir Putin had to flee the country. Others modified their business model in line with the Kremlin's guidelines (Hoffman, 2003, pp. 383–384).

When it comes to the distribution of oil and natural gas exploration rights and major contracts, several influential professional organizations lobby for the Russian energy sector at the domestic and international levels: among them, the Energy Committee at the Russian Union of Industrialists and Entrepreneurs, the Russian Fuel Union, and the Russian Gas Society (Makhortov & Tolstykh, 2008, p. 59). These groups promote their vision of Russia's energy future (Yazev, 2010).

One such group – the Russian Union of Oil and Gas Producers – serves as a liaison between the market forces and the government control of the oil and gas sector (Shmal, 2009, p. 22). In 2009, the Union's president Gennadiy Shmal stressed that the organization's main focus was to increase investments in the oil and gas industry up to a minimum of $40 billion per year. He added that just to maintain the industry's output at the current level, Russian producers use costly water injections in oil wells, negatively affecting the quality of oil. As a result, the extracted liquid contains up to 80 per cent water that needs to be separated from oil. The efficiency of this outdated extraction technology is very low: the percentage of pure oil separated from water is only 28 per cent. Shmal emphasized that Russia's current level of oil production is maintained only thanks to the country's vast reserves and new oil fields, and despite the old methods of oil extraction. In addition to that, new production sites in the harsh climate conditions of East Siberia have no necessary infrastructure and require substantial investment. In order to improve this situation for domestic producers, the Russian Union of Oil and Gas Producers has lobbied for such measures as reduced taxation, government investments, new favorable laws and regulations, the national project status, and other preferential treatment from the government (Shmal, 2009, pp. 22–24).

Discussing global oil trends and Russia's place in them, another official from the Russian Union of Oil and Gas Producers, chairman Yury Shafranik, asserts that the global era of cheap oil and easy profits for energy businesses is over because new fields require more expensive technology for oil extraction, and consumers dominate energy markets and determine prices (Shafranik, 2009, p. 26). This trend presents a significant challenge for Russia's oil and gas producers that use the Union for domestic lobbying and negotiating with government agencies that regulate various aspects of energy exploration and trade (Makhortov & Tolstykh, 2008). Such agencies include:

- The Government Commission on the Fuel and Energy Complex and Regeneration of the Mineral and Raw Materials Base (with Igor Sechin in charge)
- President's Expert Group (Arkadiy Dvorkovich)
- Presidential State-Legal Directorate (Larisa Igorevna Brychyova)
- Presidential Domestic Policy Directorate (Oleg Markovich Govorun)
- Government Commission on Legislative Activities (Sergei Sobyanin)
- Ministry of Finance (Alexei Kudrin)
- Government Department of Branch Development (Olga Pushkaryova)

- Ministry of Natural Resources and Environmental Protection of the Russian Federation (Yury Trutnev)
- Ministry of Energy (Sergei Shmatko)
- Federal Anti-Monopoly Service (Igor Artemyev)
- Federal Taxation Service (Mikhail Mokretsov and Mikhail Mishustin)
- Federal Service for the Oversight of Natural Resources (Vladimir Kirillov)
- Federal Agency on Mineral Resources (Anatoly Ledovskikh)
- Federal Tariff Service (Sergei Novikov)
- State Duma Committee for Natural Resources, Environmental Management and Ecology (Natalia Komarova)
- State Duma Committee for Budget and Taxes (Yuri Vasiliev)
- State Duma Committee for Energy (Yuri Lipatov)
- The Council of the Federation Committee on Natural Resources and Environmental Protection (Victor Orlov)
- The Council of the Federation Committee on Economic Policy, Business and Ownership (Makhortov & Tolstykh, 2008).

Other important ministries lobbied by the oil and gas companies include the Federal Energy Commission (*Federalnaya Energeticheskaya Komissiya*), which regulates and oversees the energy sector, including nuclear power, oil and gas, and utilities. Foreign policies are regulated by the Ministry of Foreign Affairs (*Ministerstvo Inostrannikh Del*), and the environmental aspects of energy resource extraction and transportation are regulated by the Ministry of Natural Resources and Ecology (*Ministerstvo Prirodnikh Resursov i Ecologii*), which is also responsible for an environmental assessment of resource projects (Newell, 2004, p. xvi).

In addition to the concerns about the cost of production and declining oil and natural gas reserves, Russian energy sector representatives have their own concerns about foreign policy and international competition. In 2010, at the Russian Petroleum and Gas Congress in Moscow, they discussed the inevitable competition and struggle for global resources and predicted that the United States would try to keep its presence everywhere around the globe, even using force.

One speaker at the 2010 Congress, Rosneft's vice president Sergey Kudryashov outlined the main tasks for Russia's immediate energy policies. Among the priorities, he mentioned Rosneft's exploration of the Vankor field (Krasnoyarsk Krai in Eastern Siberia), the Sakhalin-2 project (an LNG plant), the ESPO pipeline, better oil processing technology, Nord Stream and South Stream pipelines, the Sakhalin–Khabarovsk–Vladivostok gas pipeline, as well as deals with China. He mentioned that new transportation routes for oil and gas resources should be top priority, as well as an increase of existing projects' efficiency and more favorable taxation. To successfully develop the industry, Kudryashov urged other oil companies to focus on maximizing economic efficiency, building new infrastructure, diversifying the consumers, finding new exploration sites, and developing new products. A representative from another Russian company, Transneft, agreed with those goals and emphasized the importance of oil pipeline systems, as a means to diversify the oil routes, including the ESPO pipeline.

Unlike the Union of Oil and Gas Producers, which promotes the interests of Russian companies, other associations lobby for all energy producers. One of them, the Moscow International Petroleum Club (MMNK), was started in 1994 under the auspices of the Gore-Chernomyrdin Commission and the Russian Duma–U.S. Congress Energy Working Group (Bagirov, 2010). The club has been recognized by the United Nations as an expert group for the Economic Commission for Europe and now includes more than two dozen energy companies based in Russia, Europe, and the United States (e.g., ExxonMobil, ChevronTexaco, Shell, Total, E.ON Ruhrgas, with CNPC and Sinopec as observers) (MIPC, 2006). Co-chairmen of the club at different times have been Sergey Bogdanchikov of Rosneft, Ben Haynes of ExxonMobil, Vagit Alekperov of LUKoil, Rein Tamboezer of Shell, Alexei Miller of Gazprom, and Menno Grouvel of TotalFinaElf (MIPC, 2006). The priorities of the club have evolved with the changes in the Russian economic climate, affecting energy business (Bagirov, 2010). Since members of the club are essentially competitors, the organization has focused on issues beneficial for all of them, such as export legislation, reduction in taxes and royalties, and networking with Russian businesses (Bagirov, 2010). When Russian companies go to foreign markets, they try to influence host governments by lobbying. In the United States, however, Russian lobbies have been very limited: some companies and organizations, such as LUKoil, Itera, the Foundation for Russian American Economic Cooperation, the Russian American Pacific Partnership, and the U.S.–Russia World Forum (led by Edward Lozansky) have been trying to play the role of lobbies.

Eduard Lozansky, a major expert on Russian lobbying in the United States, stated in 2006 that 'there is no pro-Russian lobby in Washington, DC' (Verlin, 2006). To fill that void, since 1981, he has annually organized the World Russia Forum, which brings together top government officials, academics, and businessmen interested in U.S.–Russia relations. Support of the forum has come traditionally from both republican and democrat members of Congress, as the republican and democrat positions on Russia have been historically similar.

To sum up, the energy industry decision-makers in Russia are mostly the same personalities who are responsible for foreign policy. Normally, they are united around the Russian President and promote the government line.

American lobbies in Russia

Energy companies usually position themselves as pragmatic entities that follow profit opportunities and available natural resources around the world, using their governments' guidelines and informational support. In Russia, foreign companies join international business associations, such as the American Chamber of Commerce or the European Business Association (Makhortov & Tolstykh, 2008, p. 39). In addition to that, each corporation usually has a government relations department that directly deals with the host government, partners, and trade unions. Nevertheless, it has been increasingly difficult for international companies to join projects in strategically important Russian industries (Donohue & Shokhin, 2008).

One area in which most changes have occured is an opportunity for companies to sign PSAs with the Russian government. PSAs were introduced in Russia in the 1990s, when oil prices were low ($20–$30 per barrel) and the government badly needed foreign capital and expertise, as a way to attract foreign investments and to increase oil and gas production without additional government spending. PSAs, promoted by the Russian International Tax and Investment Center (ITIC) were supposed to have the status above state laws and ensure a constant level of taxation (Darmin, 2010, p. 31).

However, when oil prices soared to $100 per barrel and higher, the Russian government started to consolidate control over major oil and gas companies, so that 'the whole nation gets benefits, not just a small group of oligarchs'. Under favorable market conditions and high oil revenues, Russia also started to quickly accumulate expertise and capital (Bagirov, 2010). As a result, a new rule stipulates that the Russian state control in joint ventures has to be no less than 51 per cent. More often than not, Western oil companies in Russia now have to accept the role of consultants and service providers, rather than co-owners of projects (Darmin, 2010, p. 31).

After the initial shock of those changes in PSAs, multinationals have adjusted to the changes in the Russian business climate, and to the lack of PSAs in particular. Their lobbying organizations, such as the Petroleum Advisory Forum (PAF) started to reshape their strategies.

During an amicable period of the Obama–Medvedev 'reset' in 2009, the U.S.–Russia business relations temporarily benefited from the warming in political conditions (Obama, 2010). American companies used extensive informational support by the U.S. Embassy in Moscow and the U.S. Department of Energy, whose missions have been to 'advance the national, economic, and energy security of the United States' (Embassy, 2010). The American Chamber of Commerce (AmCham) also provided informational and lobbying support to American companies: 'Through AmCham's effective advocacy and high credibility, member companies are assured of having their concerns heard by decision-makers in both Washington and Moscow' (AmCham, 2010). AmCham's Energy Committee, co-chaired by the president of Chevron Neftegaz, Darrell Cordry was helping American businesses secure contracts in Russia (AmCham Energy Committee, 2010).

AmCham was an active proponent of Russia's joining WTO (Kostyaev, 2011). Previously, the U.S.–Russia trade and investment had been negatively affected by Russia's lack of WTO membership, which Russia acquired only on August 22, 2012, as the WTO's 156th member. One of the conditions set by Western countries for Russia to join WTO has been the liberalization of Russia's domestic energy prices and the establishment of free trade institutions. Potentially, Russia could utilize WTO functions – as the conference organization for multilateral trade issues; the depositary of binding international trade conventions; and conflict resolution, arbitration, and trade dispute mechanisms – to diversify the country's resource-based economy and move towards knowledge-based development (Aslund, 2010).

American law firms also have lobbying activities in Russia: such firms as Akin Gump Strauss Hauer & Help; Hogan & Hartson; Cassidy & Associates; and Fleishman-Hillard practice lobbying and public relations on the Russian market (Kostyaev, 2011).

Sometimes, however, it has been unfavorable for a Russian company to have too close ties or deals with American partners – that is what happened to Russia's Yukos that has become a prominent case of the struggle between an independent oligarch and the Russian government.

The Yukos case

The case of Yukos's rise and demise is a culminating example of Russia's overall economic development in stages the 1990s, partly due to the dysfunctional tax system (from which Yukos benefited via low taxation of its revenues), and the strengthening of state control during the first two terms of Putin's presidency. The Yukos trial is the most prominent case in which economic, political, and geo-strategic aspirations of an oil company and its leader clashed with the government (Frye, 2010, p. 190).

The Loans-for-Shares program was implemented in the mid-1990s by the Russian government, which badly needed financial resources to cover the state budget deficit. During the program, major state enterprises were leased out in exchange for monetary loans in the process of auctions, and the access to auctions was limited to an internal circle of individuals. The program implied that if the government did not repay the loans to the banks, the enterprises would be transferred to the banks at a fraction of their real costs. That is exactly what happened to a number of major state assets, including what later became Yukos (Frye, 2010).

Emerging from the Loans-for-Shares program of the 1990s and a number of consolidations, by the year 2000 Yukos became the biggest and the richest oil company in Russia. Its owner Mikhail Khodorkovsky felt very confident not only in his business, but also in de-facto foreign policy making, as he was signing contracts directly with foreign companies and foreign governments for oil and gas trade. He was becoming a real political force, challenging Vladimir Putin's consolidation of the state power and re-nationalization tendencies in the Russian energy sector during Putin's first term.

In 2003, Mikhail Khodorkovsky visited the United States and presented Russia as a supplier of energy resources that would provide the West with an alternative supply of oil, different from the unstable Middle Eastern oil supply (Hoffman, 2003, p. 382). By 2004, Yukos became Russia's largest oil producer, exported oil by tankers to the United States and had plans to sign a 20-year oil delivery contract with China. Khodorkovsky was acting independently and pursuing his own deals while President Putin was strengthening the government's control over the Russian upstream. Yukos's opposition to the government worsened when Khodorkovsky decided to sell a big share of the company's assets to ExxonMobil and Chevron (Mankoff, 2009, p. 25). The Russian government became concerned about Yukos's plans to sell a part of the company to American oil majors.

In a way, Yukos was acting as a sovereign power and making foreign policy, rather than following the government's instructions. Yukos signed a protocol of understanding with ExxonMobil about the purchase of shares, without consulting with Vladimir Putin. To add to the Russian government's concerns, Khodorkovsky appointed American executives as Yukos's top employees: Bruce Misamore of Marathon Oil and PennzEnergy as Chief Financial Officer and Steven Theede of ConocoPhillips as his Chief Operating Officer. Other foreign specialists – Sarah Carey, Rai Kumar Gupta of Phillips, Bernard Loze of France, Jacques Kosciusko-Morizet of Credit Lyonnais, and Michel Soublin of Schlumberger – joined Yukos's board of directors. Strengthening his ties with the United States, Khodorkovsky set up a philanthropic foundation and provided grants and sponsorship to the U.S. Library of Congress. Shortly after, Khodorkovsky was arrested on tax evasion charges (Goldman, 2008, p. 123).

Yukos was charged with tax fraud and ordered to pay back millions of dollars to the Russian federal budget. The company's assets were frozen and eventually were purchased by Russia's state-controlled company Rosneft (Volkov, 2008, p. 241). Foreign companies were not allowed to buy parts of Yukos (Mankoff, 2009, p. 37). Some scholars compared what happened to Yukos to the American conglomerate Standard Oil, which had been broken apart in 1911 because it became too powerful. Both Yukos and Standard Oil came into existence during the 'wild capitalism' epoch in their respective countries. Similar to Theodore Roosevelt, Vladimir Putin was trying to limit the power of leading tycoons (Volkov, 2008, p. 241).

Mikhail Khodorkovsky was put in jail and stayed there for almost a decade. Human rights activists, such as Khodorkovsky's lawyer Robert Amsterdam, worked to bring international awareness of this case. Khodorkovsky was released from jail unexpectedly in December 2013, when he was pardoned by Vladimir Putin. After leaving the prison, Khodorkovsky arrived in Germany and since then has not come back to Russia. By that time, as Western media noted, his image had been reinvented from a 'powerfully wealthy, often arrogant oligarch' into 'a respected dissident, political thinker and editorial writer who argued for social justice' ('Russian Tycoon Mikhail Khodorkovsky Freed from Prison, Flies to Germany', 2013).

As Yukos disintegrated, another Russian oligarch, Roman Abramovich, was also planning to sell parts of his company, Sibneft, to Chevron-Texaco, Shell, and Total. Similar to Yukos, Sibneft was quickly charged with tax fraud. To avoid criminal prosecution, Abramovich quickly sold his shares in Sibneft to Gazprom, and a new business entity – Gazpromneft – was formed out of this deal. After this deal, the Russian state's ownership share reached 30 per cent of Russia's overall oil production (Goldman, 2008, p. 123).

Another energy company with tight connections to the United States – ITERA – was the second largest Russian producer of natural gas in the 1990s, headquartered in Jacksonville, Florida. In the 1990s, ITERA, led by Igor Makarov, started to deliver natural gas from Turkmenistan to Ukraine via Gazprom's pipeline system. During the Loans-for-Shares initiative, ITERA acquired substantial assets from Gazprom at a sub-market cost. When Rem Vyakhirev, the head of Gazprom who was favorable to ITERA, lost his position as Gazprom's CEO, ITERA had to return Gazprom assets and also lost the Turkmenistan-Ukraine gas trade opportunity in 2002 (Goldman, 2008).

Conclusion: the past and the future of U.S.–Russia relations

The case of U.S.–Russia is a multi-faceted case that requires a multi-aspect explanation. Russia, as a large oil producer in the world (competing with Saudi Arabia), and the United States, the largest oil consumer, could potentially increase oil and gas trade and investments. Initally, the early 1990s were an optimistic start, when Russia opened its doors to American advisors and businesses. President Clinton and his team (Warren Christopher, Madeleine Albright, Les Aspin, William Perry, Colin Powell, Anthony Lake, and Strobe Talbott) promoted economic liberalization as Russia's future course of development (MacLean, 2006, p. 4):

> … the United States treated Russia as a kind of pet project for the contemporaneous creation of capitalism and democracy, in which American lawyers, academics, and newly minted MBAs descended on Moscow to rewrite tax laws, design institutions, and peddle advice.
>
> (Appel, 2008)

During that stage, the economic potential of cooperation was considered very high by both governments, and they encouraged businesses to open joint ventures and sign production sharing agreements. Despite the chaos and instability of the Russian economy, the American government focused on opportunities presented by a newly open Russia. Major energy companies followed suit.

The economic crisis of 1998, caused by both domestic (inflation, tax evasion, corruption) and international (the Asian crisis contagion) factors, became a watershed that brought Vladimir Putin to power and created favorable conditions for his strengthening of state power and Russia's international stance. The early years of the Putin administration and the George W. Bush administration were relatively cooperative and productive, especially after the 9/11 terrorist attack when Russia opened its air space and territory for U.S. military cargoes and shared intelligence. However, later the U.S.–Russia relations nevertheless started to quickly deteriorate, plagued by such issues as NATO expansion, Iran, U.S. missile defense and especially by the Iraq War of 2003, which Russia strongly opposed. The lowest point of the relations between the two countries was reached during the Russia–Georgia War of 2008, when the United States took Georgia's side and some American politicians even suggested that the United States should start a military intervention against Russia.

As a result, the last decade was a period when the share of American investments in the Russian economy, and in the energy sector in particular, fell drastically. For almost a decade, major U.S. companies (Chevron, ExxonMobil, ConocoPhillips) could not join new projects and were struggling to keep the exisisting projects (in Sakhalin, Shtokman, the Black Sea, and others) going. The Russian government reversed the Westernizing trend of the 1990s and increased state ownership and control over Russia's energy resources. Oil and gas became not only a source of revenue for the state budget, but also a tool of foreign policy, geopolitical advantage, and Russia's national pride.

References

Allison, R. O. Y. (2013). Russia and Syria: Explaining Alignment with a Regime in Crisis. *International Affairs*, 89(4), 795–823. doi: http://dx.doi.org/10.1111/1468-2346.12046

AmCham. (2010). About Us. Retrieved September 12, 2010, from http://www.amcham.ru/about

AmCham Energy Committee. (2010). Energy. Retrieved September 12, 2010, from http://www.amcham.ru/eng/committees/energy

Anishyuk, A. (2009, December 14). LUKoil Snaps Up Coveted Iraqi Field. *The Moscow Times*, pp. 1–2.

Appel, H. (2008). Is It Putin or Is It Oil? Explaining Russia's Fiscal Recovery. *Post-Soviet Affairs*, 24(4), 301–323. doi: 10.2747/1060-586X.24.4.301

Arai, H. (2012). Foreign Direct Investment of Russia. Retrieved June 10, 2013, from http://www.ide.go.jp/English/Publish/Download/Apec/1863ra0000006alg-att/1863ra0000006d5s.pdf

Aslund, A. (2010). Why Doesn't Russia Join the WTO? *The Washington Quarterly*, 33(2), 49–63.

Bagirov, T. (2010). Media Center – Interviews. *RussiaEnergy.Com*. Retrieved September 22, 2010, from http://www.russiaenergy.com/index.php#state=InterviewDetail&id=582

Bazhanov, Y. (2010). Why Russia Needs a Strong NATO. *The Moscow Times*, (22 September, 2010). Retrieved from http://www.themoscowtimes.com/opinion/article/why-russia-needs-a-strong-nato/416788.html

Besedovskiy, R. (2010). East Siberian Crude: Closer to California Refineries. *Oil and Gas Eurasia*, 7 (July–August 2010). Retrieved from http://www.oilandgaseurasia.com/articles/p/123/article/1282/

Blanchard, R. D. (2005). *The Future of Global Oil Production: Facts, Figures, Trends and Projections, by Region*. Jefferson, NC: McFarland & Company, Inc.

Chevron. (2010). Russia Fact Sheet. Retrieved from www.chevron.com/Documents/Pdf/RussiaFactSheet.pdf

Chevron Concern About Russian Law on Foreign Investment and its Impact on Oil Exploration. (2010). *Neftegas.Ru*. Retrieved from http://www.neftegaz.ru/en/news/view/93783

Chevron, Gazprom Identify Possible Joint Projects. (2009). *Bloomberg, December* 8. Retrieved from http://www.chron.com/disp/story.mpl/business/energy/6758735.html

Chevron i Gazprom Neft Obsuzhdayut Proekti. (2009, December 9). *Kommersant*, p. 11.

Chevron Neftegaz. Inc. (2000). Retrieved April 12, 2010, from http://www.mmnk.org/ChevInfoFeb2000.html

Chevron Signs Project in Russia. (2010). *Daily Markets* (June 21). Retrieved from http://www.dailymarkets.com/stock/2010/06/21/chevron-signs-project-in-russia/

Cohen, S. (2001). *Failed Crusade: America and the Tragedy of Post-Communist Russia*. New York, NY: W.W. Norton & Company.

Colton, T. (2008). *Yeltsin: A Life*. New York, NY: Basic Books.

Companies & Markets; Rosneft Finishes Buying TNK-BP, Becomes World's Largest Publicly-traded Oil Co. (2013, March 21). *Interfax : Russia & CIS Business & Financial Daily*.

Cooper, W. H. (2014). International Trade and Finance: Key Policy Issues for the 113th Congress, Second Session. *Current Politics and Economics of the United States, Canada and Mexico, 16*(3), 383–424.

Darmin, A. (2010). Investitsii: Pravila Investirovaniya v Rossiyu. [Investments: The Rules of Investing in Russia]. *Neftegas.Ru*, 5, 30–37.

Desai, R. M., & Goldberg, I. (2008). Introduction. In R. M. Desai & I. Goldberg (Eds.), *Can Russia Compete?* (pp. 1–11). Washington, DC: The Brookings Institution Press.

Doing Business in the New Independent States – A Resource Guide. (1994) (July 1994 ed., Vol. 5, pp. 497–511). U.S. Department of State Dispatch.

Donaldson, R., & Donaldson, J. (2003). The Arms Trade in Russian-Chinese Relations: Identity, Domestic Politics, and Geopolitical Positioning. *International Studies Quarterly*, 47, 709–732.

Donohue, T., & Shokhin, A. (2008). U.S.-Russia Business Dialogue. *RIA Novosti*, (July 28, 2008). Retrieved from http://rbth.ru/articles/2008/07/28/Business_dialogue.html

Dougher, R. (2009). America's Oil and Natural Gas History: Securing America's Energy Future – A Reality Check.

EIA. (2010). Russia: Country Analysis Brief. Retrieved from www.eia.doe.gov

English, R. (2000). *Russia and the Idea of the West: Gorbachev, Intellectuals, and the End of the Cold War*. New York, NY: Columbia University Press.

Foglesong, D. S. (2007). *The American Mission and the 'Evil Empire': The Crusade for a 'Free Russia' since 1881*. Cambridge, MA: Cambridge University Press.

Friedman, J. (2011). Russia's Nuclear Forces and Doctrine. *CSIS*. Retrieved from https://csis.org/blog/russias-strategic-nuclear-forces-and-doctrine

Frye, T. (2000). *Brokers and Bureaucrats: Building Market Institutions in Russia*. Ann Arbor, MI: The University of Michigan Press.

Frye, T. (2010). *Building States and Markets After Communism: The Perils of Polarized Democracy*. New York, NY: Cambridge University Press.

FSGS. (2015). Interactivnaya Vitrina. Retrieved March 19, 2015, from http://cbsd.gks.ru/

Gaddy, C. (2006). The U.S. and Soviet Oil Production, 1945-1990: Brookings Institution.

Gati, T. T., & Christiansen, T. L. (2003). The Political Dynamic. In J. H. Kalicki & E. K. Lawson (Eds.), *Russian-Eurasian Renaissance? U.S. Trade and Investment in Russia and Eurasia* (pp. 447–459). Washington, DC: Woodrow Wilson Center Press.

Gimbel, B. (2009). Russia's King of Crude. *Fortune*, 159, 88–92.

Goldgeier, J. M., & McFaul, M. (2005). What to Do about Russia. *Policy Review*, 133, 45–62.

Goldman, M. (2008). *Petrostate: Putin, Power, and the New Russia*. Oxford: Oxford University Press.

Guseinov, V. (2008). *Politica u Poroga Tvoego Doma. Izbrannie Publitsisticheskie i Analiticheskie Statji*. Moscow: Krasnaya Zvezda.

Hoffman, D. E. (2003). Oligarchic Capitalism in Russia. In J. H. Kalicki & E. K. Lawson (Eds.), *Russian-Eurasian Renaissance? U.S. Trade and Investment in Russia and Eurasia* (pp. 371–387). Washington, DC: Woodrow Wilson Center Press.

Hueper, P. F. (2003). The Energy Locomotive. In J. H. Kalicki & E. K. Lawson (Eds.), *Russian-Eurasian Renaissance? U.S. Trade and Investment in Russia and Eurasia* (pp. 173–195). Washington, DC: Woodrow Wilson Center Press.

Iskyan, K. (2012). Russia's 18-Year Odyssey To Join The WTO Finally At An End. *Global Finance*, *26*(2), 8.

Ivanitskaya, E. V. (2009). Moskva – Epitsentr Mezhdunarodnikh Neftegazovikh Meropriyatiy. [Moscow: The Epicenter of International Oil and Gas Events]. *Truboprovodniy Transport: Teoriya i Praktika*, *4*(16), 9–11.

Johansson, K. (2008). Foreign Direct Investment in Russia: A survey of CEOs. Moscow: Foreign Investment Advisory Council.

Kalicki, J. H., & Lawson, E. K. (Eds.). (2003). *Russian-Eurasian Renaissance? U.S. Trade and Investment in Russia and Eurasia*. Washington, DC: Woodraw Wilson Center Press.

Kostyaev, S. (2011). Americanskiy Lobbyism v Rossii. *Lobbying.Ru*. Retrieved from http://www.lobbying.ru/print.php?article_id=6963

Kramer, A. E. (2009, February 18). Russia, Looking Eastward, Opens a Gas Plant to Supply Asian Markets. *The New York Times*, p. B3. Retrieved from http://www.nytimes.com/2009/02/19/business/worldbusiness/19ruble.html?emc=eta1

Krauss, C. (2015, January 12). Oil Prices: What's Behind the Drop? Simple Economics. *The New York Times*.

The Kremlin has Nothing to Fear from Left-wing Opposition. (2014, December 23). *OpenDemocracy*.

Lieven, A. (2000). Against Russophobia. *World Policy Journal*, *XVII*(4).

Lotarski, S. S. (2003). Expanding U.S.-Russia Trade. In J. H. Kalicki & E. K. Lawson (Eds.), *Russian-Eurasian Renaissance? U.S. Trade and Investment in Russia and Eurasia* (pp. 29–41). Washington, DC: Woodrow Wilson Center Press.

Maass, P. (2009). *Crude World*. New York, NY: Alfred A. Knopf.

MacLean, G. A. (2006). *Clinton's Foreign Policy in Russia: From Deterrence and Isolation to Democratization and Engagement*. Hampshire: Ashgate.

Makhortov, E., & Tolstykh, P. (2008). *Otraslevoye Lobbirovanie v Rossiiskoi Federatsii na Primere Neftegazovogo Kompleksa*. Moscow: GR Research and Consulting Center: Lobbying.Ru.

Malik, N., & Henshall, A. (2010, July 6). Russia-to-Asia Pipeline Takes Detour to U.S. *Wall Street Journal*.

Mankoff, J. (2009). *Russian Foreign Policy: The Return of Great Power Politics*. Lanham, MD: Rowman & Littlefield, Inc.

Mann, S. R. (2003). Caspian Futures. In J. H. Kalicki & E. K. Lawson (Eds.), *Russian-Eurasion Renaissance? U.S. Trade and Investment in Russia and Eurasia* (pp. 147–171). Washington, DC: Woodrow Wilson Center Press.

Marshall, Z. B. (2003). Russia's Investment Outlook. In J. H. Kalicki & E. K. Lawson (Eds.), *Russian-Eurasian Renaissance? U.S. Trade and Investment in Russia and Eurasia* (pp. 9–28). Washington, DC: Woodrow Wilson Center Press.

Massie, S. (2008). *Reagan's Evolving Views on Russia and Their Relevance to Today*. Washington, DC: Woodrow Wilson International Center for Scholars.

Matlock, J. F. (2010). *Superpower Illusions: How Myths and False Ideologies Led America Astray – And How to Return to Reality*. New Haven, CT: Yale University Press.

Melville, A., & Shakleina, T. (2005). Introduction. In A. Melville & T. Shakleina (Eds.), *Russian Foreign Policy in Transition* (pp. ix–xiii). Budapest: Central European University Press.

MIPC. (2006). Moscow International Petroleum Club. Retrieved September 21, 2010, from http://www.mmnk.org/main.htm

Mortished, C. (2008, September 3). Europe Has Weak Hand in a Poker Game of Power. *The Times (London)*, p. 49.

Newell, J. (2004). *The Russian Far East: A Reference Guide for Conservation and Development*. McKinleyville, CA: Daniel & Daniel.

Novelli, C. A. (2003). Trade Agreements and the WTO. In J. H. Kalicki & E. K. Lawson (Eds.), *Russian-Eurasian Renaissance? U.S. Trade and Investment in Russia and Eurasia* (pp. 429–446). Washington, DC: Woodrow Wilson Center Press.

Obama, B. (2010). Remarks by the President on the Announcement of New START Treaty. Retrieved from http://www.whitehouse.gov/the-press-office/remarks-president-announcement-new-start-treaty

Parker, J. W. (2012). *Russia and the Iranian Nuclear Program: Replay or Breakthrough?* Washington, DC: National Defense University Press.

Piadyshev, B. (2006). Russia's Priorities. *International Affairs*, 52(4), 8–49.

Projects in Russia. (2010, November 10). Retrieved from http://www.conocophillips.ru/EN/project-russia/Pages/index.aspx

Quester, G. H. (2007). Energy Dependence and Political Power: Some Paradoxes. *Democratizatsiya*, 15(4), 445–453.

Reuters. (2011). REG – Rosneft – Publication of Rosneft's Annual Report for 2010. Retrieved September 8, 2015, from http://www.reuters.com/article/2011/04/28/idUS219353+28-Apr-2011+RNS20110428

Rosneft Teams Up with ExxonMobil in Arctic Deal. (2011). *Business Today*. Retrieved from http://www.businesstoday-eg.com/business/asia-pacific/rosneft-teams-up-with-exxon-mobil-in-arctic-deal.html

Russia. (2010, September 20). *Russia & CIS Business & Financial Daily*. Retrieved from http://www.lexisnexis.com.libproxy.usc.edu/hottopics/lnacademic/?

Russia Armaments. (2015). Mount Albert: Acquisdata Pty Ltd.

Russia-Sakhalin-1. (2010). ExxonMobil.

Russia's Putin Signs Law on 'Undesirable'; Foreign NGOs. (2015, May 23). *BBC Monitoring Newsfile*.

Russian foreign minister interviewed on EU sanctions, Ukraine, Kuril Islands. (2015, May 22). *BBC Monitoring Former Soviet Union*.

Russian President Calls for Improving Investment Climate. (2010, February 2). *BBC Monitoring Former Soviet Union*.

Russian Tycoon Mikhail Khodorkovsky Freed from Prison, Flies to Germany. (2013, December 20). *Oakland Tribune*.

Russian–American Business Dialogue. (2003). Retrieved September 20, 2010, from http://moscow.usembassy.gov/fact_09272003d.html

RussiaProfile.org. (2009). Production Sharing Agreements in Sakhalin. Retrieved from http://www.russiaprofile.org/resources/territory/districts/sakhalin/economy

Ryabova, I. (2010). Medlenno i Neverno: ConocoPhillips Objasnila Prichini Prodazhi Aktsii LUKoila. *Lenta.Ru*, (March 29, 2010). Retrieved from http://www.lenta.ru/articles/2010/03/29/lukoil/

Seib, C. (2008, September 9). Russian Roulette is a Risky Game to Play, But the Rewards Can Be Huge. *The Times (London)*, p. 61.

Shaffer, M., & Kuznetsov, B. (2008). Productivity. In R. M. Desai & I. Goldberg (Eds.), *Can Russia Compete?* (pp. 12–34). Washington, DC: The Brookings Institution Press.

Shafranik, Y. (2009). Globalnaya Energetika i Rossiya. [The Global Energy Sector and Russia]. *Vestnik Aktualnikh Prognozov. Rossiya: Tretye Tisyacheletije, 21*, 25–27.

Shmal, G. (2009). Optimizma Teryat 'Ne Sleduyet, No Kompleksniye Programmi Razvitiya Nado Razrabativat' Seichas. [We Should Not Lose Optimism But We Need to Develop Complex Development Programs]. *Vestnik Aktualnikh Prognozov. Rossiya: Tretye Tisyacheletije, 21*, 22–23.

Simonov, K. (2011). Chro Stoit Za Sdelkoi Mezhdu 'Gazpromom' i Nemetskoi Wintershall. *Forbes.ru*

Socor, V. (2010). La France D'Abord: Paris First to Capitalize on Russian Military Modernization. *Eurasia Daily Monitor, 7*(29). Retrieved from http://www.jamestown.org/single/?no_cache=1&tx_ttnews[tt_news]=36029&tx_ttnews[backPid]=13&cHash=2426a5184c

Starr, F. (2003). The Investment Climate in Central Asia and the Caucasus. In J. H. Kalicki & E. K. Lawson (Eds.), *Russian-Eurasian Renaissance? U.S. Trade and Investment in Russia and Eurasia* (pp. 73–91). Washington, DC: Woodrow Wilson Center Press.

Stern, J. (2005). *The Future of Russian Gas and Gazprom*. Oxford: The Oxford University Press.

Sueldo, A. (2011, July 19). Holding the Reset Hostage. *The Moscow Times*. Retrieved from http://www.themoscowtimes.com/opinion/article/holding-the-reset-hostage/440746.html#axzz1SZh9ZuMu

Talbott, S., & Montes-Negret, F. (2008). Foreword. In R. M. Desai & I. Goldberg (Eds.), *Can Russia Compete?* (pp. vii–xi). Washington, DC: Brookings Institution Press.

Taylor, J. G. (2010). *Sakhalin-1: Prosperity Through Partnership. Presentation at the Russian-American Pacific Partnership Meeting*. Presentation. Exxon Neftegaz Limited. Portland, OR.

Trenin, D. (2002). *The End of Eurasia: Russia on the Border Between Geopolitics and Globalization*. Washington, DC: Carnegie Endowment for International Peace.

Tsygankov, A. (2005). Vladimir Putin's Vision of Russia as a Normal Great Power. *Post-Soviet Affairs, 21*(2), 132–158.

Tutushkin, A., & Vasilyev, I. (2009). Obeschannogo 12 Let Zhdut. *Vedomosti*, p. B02.

U.S. Commercial Services in Russia. (2009). Retrieved April 8, 2009, from http://www.exim.gov/russia/staff.cfm

U.S. Embassy (2010). U.S. Department of Energy, Moscow Office. Retrieved September 12, 2010, from http://moscow.usembassy.gov/energy.html

United States: Shale Gas Boom to Boost Industry, Trade. (2013). Oxford: Oxford Analytica Ltd.

Verlin, Y. (2006). Edward Lozansky: 'Prorossiiskogo Lobby v Washingtone Net'. *Profile, 15*(477). Retrieved from http://www.profile.ru/items/?item=18812

Volkov, V. (2008). Standard Oil and Yukos in the Context of Early Capitalism in the United States and Russia. *Democratizatsiya, 16*(3), 240–264.

Wadhwa, V. (2010). Can Russia Build A Silicon Valley? *Tech Crunch*. Retrieved from http://techcrunch.com/2010/09/12/can-russia-build-a-silicon-valley/

Watkins, E. (2009). Putin Launches First Phase of ESPO Oil Line. *Oil and Gas Journal*, (December 29).

Yazev, V. (2010). *Russia and the International Energy Cooperation in the XXI Century*. Moscow: Granica.

Yergin, D. (2011). *The Quest: Energy, Security, and the Remaking of the Modern World*. New York: Penguin Books.

Zak, D. (2015, May 23). U.N. Nonproliferation Conference Collapses Over Mideast Issue. *The Washington Post*.

3 U.S.–Azerbaijan energy diplomacy

BTC and beyond

Introduction

Azerbaijan, an oil-rich country with a population of 9.5 million and a GDP per capita of about $8,000, has long been an interest for Western energy companies ('Azerbaijan: Country Fact Sheet', 2015). Western companies invested in multi-billion dollar projects in the Caspian region – the Baku–Tbilisi–Ceyhan (BTC) pipeline, the Shah-Deniz natural gas field, the Baku-Supsa pipeline, and others. The U.S. government has consistently provided strong diplomatic support to Azerbaijan, and the two countries have successfully cooperated despite some economic and geographical challenges and obstacles.

First, almost no oil is exported directly from Azerbaijan to the United States. Azerbaijan is practically landlocked from major oceans and has to rely on transit countries for oil and gas transportation. Azerbaijan's oil reserves are not as big as Russia's or Saudi Arabia's. Experts even disagree about the amount of oil and gas reserves that Azerbaijan possesses, and whether Azerbaijan will be able to fill the multi-billion-dollar pipelines to capacity.

Second, Azerbaijan, similar to Russia, has a strong leader, President Ilham Aliyev. His father, Heydar Aliyev, was the party leader of Soviet Azerbaijan and then became Azerbaijan's president after the dissolution of the USSR. Ilham Aliyev was prime minister at the time and became president of Azerbaijan after his father's death. Human rights non-governmental organizations (NGOs), and even the U.S. Department of State, have been concerned with Azerbaijan's human rights situation. However, it has not slowed down the U.S. government's involvement in Azerbaijan's energy projects.

Third, since the 1980s, Azerbaijan and Armenia have been in a prolonged territorial dispute over Nagorno-Karabakh, which left hundreds of thousands of refugees and is still the focal point of both countries' foreign policies. Armenia has a powerful lobby in the United States. In addition to that, Armenia has a complicated relationship with Turkey because of the disagreements about mass killings of the Armenians by the Turkish in 1915–1923. Potentially, one might expect that the Armenian lobby would try to prevent the U.S. involvement in the BTC project, which included both Azerbaijan and Turkey.

Fourth, Azerbaijan has close ties with Iran: about 16 per cent of Iran's population is ethnically Azeri ('Azeri MP Urges More Attention to Problems of Ethnic Azeris in Iran', 2014). Despite serious disagreements between the United States and Iran on the Iranian nuclear program and other issues, Azerbaijan has been able to keep working relations with both states.

After the breakup of the Soviet Union in 1991, the post-Soviet Azerbaijan quickly emerged as an attractive point for foreign investments. Since the 1990s, major international companies poured billions of dollars into the country's projects. The 1,099-mile BTC pipeline, finished in 2006, which cost $4 billion to build, was one of the main projects. Another $25-billion international project, the Shah Deniz-II, has been in development for production and transportation of natural gas.

These multi-billion projects have been backed by the United States government and other Western countries. The United States and Europe have had a strong interest in building transportation routes (mainly pipelines) that would bypass Russia and provide an alternative source of oil and gas for Western Europe.

The political transformation of Azerbaijan in the 1990s

Azerbaijan announced its independence from the Soviet Union on October 18, 1991, followed by a national referendum on December 29, 1991. The referendum confirmed that the voters wanted to establish an independent state ('18 October – The National Independence Day', 2009). In the early 1990s, the country had a quick succession of presidents, since the political situation was still unstable. Ayaz Mutallibov, who had been the last leader of the Soviet Republic of Azerbaijan, became the first president of the independent Azerbaijan in 1990–1992. He was followed by Yagub Cavad oglu Mammadov, who stayed president only for two months (from March 6 to May 14, 1992). After Mammadov, Isa Yunus oglu Gambarov held the position from May 19 to June 16, 1992, followed by Abulfaz Gadirgulu oglu Elchibey, who remained in power for more than a year (from June 16, 1992 to September 1, 1993).

The father of the current president Ilham Aliyev, Heydar Alirza oglu Aliyev, became the next president on June 24, 1993 and stayed in power for 10 years until his death in October 2003. He was succeeded by his son Ilham Heydar oglu Aliyev, who has been president since October 31, 2003. Having already held a government position during the Soviet time, Heydar Aliyev was a prominent figure on Azerbaijan's political arena for several decades. His career started as a KGB officer in the 1940s and 1950s. In 1967, he became the head of the country's KGB, and two years later, in 1969, the First Secretary of the Communist Party's Central Committee. Aliyev held this position until 1987. After a brief service as the Chairman of the Supreme Soviet in his native Nakhchivan (an Azerbaijani province), Aliyev became President of Azerbaijan in 1993. There are various accounts of Aliyev's rise to presidency. The Heydar Aliyev Heritage Research Center stated that the nation had wanted him to 'lead the country' ('Heydar

Alirza oglu Aliyev', 2009). Other sources, such as NATO archives, describe the change in power as a military coup against his predecessor, President Abulfaz Elchibey ('Visit to Baku, Azerbaijan, Jan 22–24 October 1998', 1999).

The critics argue that the state system created by Heydar Aliyev was lacking political diversity. As Vagif Guseinov of Russia's Institute of Strategic Assessment and Analysis put it, a stable state 'machine', created by Heydar Aliyev, allowed painlessly transferring the power from the father to the son, through the prime-minister position to the presidency. Both presidents concentrated an unprecedented power and wealth in the hands of their family (Guseinov, 2008).

Heydar's son, Ilham Aliyev, held high-level positions since 1994, when he became Vice-President of SOCAR and later Prime Minister. Heydar Aliyev was terminally ill at the time and was ready to pass on his power. Two months after his appointment as Prime Minister, in 2003 Ilham Aliyev won presidential elections, replacing his father as the president. The governments of the United States, Russia, Turkey, and other states recognized the elections as fair and praised Azerbaijan for the 'successful choice of leadership', despite some protests by Azerbaijan's opposition movements (Guseinov, 2008, p. 40). Ilham Aliyev continued the policies started by his father, such as energy deals for foreign investors and economic stability.

Cooperation with the United States

Acquiring independence, the post-Soviet nation designed its national symbols and national currency. They were similar to those of the pre-Soviet Azerbaijani Republic that had existed briefly between 1918 and 1920, before it was integrated as a part of the Soviet Union (Finn, 2009, p. 100). In the early 1990s, the newly independent Azerbaijan started to establish working relations with major players in the region, such as Iran and Russia, especially that both countries had a substantial ethnic Azeri population.

After the independence, Azerbaijan's government started to reorganize its Soviet-type economy into a market-based model. The country's main industry, the oil and natural gas sector, underwent restructuring: two major state oil enterprises, Azneft and Azerneftkimya, were merged into one – the State Oil Company of Azerbaijan Republic (SOCAR). SOCAR has been the country's vertically integrated oil and gas company that explores, produces and trades energy resources (Azerbaijan.az). The company is still in the government's hands, even though it has gone through a major redistribution of assets and reformulation of business processes.

Heydar Aliyev stabilized the domestic political situation (four presidents in three years before him had been quite indicative of the previous instability) and started to revive the country's production and trade in energy resources. According to Head of Foreign Relations Department of the Aliyev Administration, Novruz Mammadov, Heydar Aliyev's predecessor Elchibey had promoted the country's 'self-isolation'. Contrary to Elchibey's policies, Aliyev prioritized the country's openness to the West while keeping productive relationships with Russia, Iran, and other neighbors (N. Mammadov, 2009, p. 150). In that, he was very successful: he found a productive balance between Moscow and Washington, ensuring lucrative oil and gas contracts and a stable inflow of petro-dollars.

For Azerbaijan, one of the most important tasks was to establish solid bilateral relations with the United States, the most powerful country after the end of the Cold War. In 1992, U.S. Secretary of State Cyrus Vance visited a newly opened U.S. embassy in Baku. Shortly after that visit, Azerbaijan was accepted into the NATO Partnership for Peace program. Azerbaijan has been a stable partner in this program, even though there was a case when this status was almost revoked from Azerbaijan by Rep. Howard Berman (D-CA) in 2012, because Azerbaijan 'pardoned axe-murderer Ramil Safarov and his promotion to the rank of major in the Azerbaijani military' ('Howard Berman Writes Secretary of State Clinton, Calls for Azerbaijan's Suspension from NATO Partnership for Peace Program and Ending Arms Sales to Azerbaijan', 2012).

U.S. diplomats witnessed the internal struggles in the newly independent state. The first Charge d'Affaires of the U.S. Embassy, Robert Finn, recalls that his job was quite challenging: Azeri nationalism was on the rise, the country was involved in a bloody war with Armenia, and whole communities were displaced because of the violence (Finn, 2009, pp. 99–102). The Armenia–Azerbaijan war had started in 1988 over the province of Nagorno-Karabakh, and the international community was trying to mediate the conflict ('Azerbaijan, Armenia Seek End to War', 1992, p. 11).

Despite Azerbaijan's conflict with Armenia, a stream of high-profile American officials kept coming to Baku, mostly to explore opportunities for energy cooperation. Among them were the U.S. Secretary of Energy, Hazel O'Leary and the Undersecretary of State, Strobe Talbott (Finn, 2009, p. 102). At the time, U.S. oil production was in decline, and the United States was toying with an idea of Azerbaijan's oil exports to the United States and other NATO partners. The demand for oil in developed countries was steadily growing, especially in the transportation sector, and such an idea seemed feasible (Sabonis-Helf, 2009). The U.S. Department of Commerce saw opportunities for American companies in the energy sector of Azerbaijan, especially in project management, engineering, drilling, and other extraction activities ('Doing Business In Azerbaijan: A Country Commercial Guide for U.S. Companies', 2008, p. 18). The former Soviet republics were considered a 'safe haven' and a new supplier of oil for the West, and new pipelines were being planned (Mortished, 2008).

The economic potential of BTC

When governments and energy companies consider opportunities for energy trade and investment abroad, an analysis usually starts with the availability of energy resources in a chosen country and its investment climate. Experts have provided conflicting estimates of Azerbaijan's oil and natural reserves, and some have even expressed concerns that the reserves might be in decline. However, the investment climate has been increasingly favorable and stable for foreign investors.

Oil and natural gas reserves and pipeline routes

Azerbaijan has undoubtedly made tremendous economic progress since it became independent in 1991. Similar to Russia, Azerbaijan went through an initial decline of its economy in the very beginning of the 1990s, right after the dissolution of the Soviet Union. The country's GDP in 1994 was only 44 per cent of the 1990 level. An armed conflict with Armenia over Nagorno-Karabakh and the internal displacement of more than half a million people created an additional burden on Azerbaijan's economy (Maharramov, 2010, p. 40). Similar to Russia, Azerbaijan's oil industry also suffered a lack of investment and modern technology.

Historically, Azerbaijan is one of the pioneers of oil production in the world: in fact, at the beginning of the twentieth century, Azerbaijan was the biggest oil producer in the world. The country's oil production started even earlier, in the mid-nineteenth century, when Azerbaijan was already one of the leading oil states. First industrial-scale oil production in Baku in 1847 was on the Ramany, Balakhany, Sabunchu, and Bibi-Heybat oil fields. According to the President of the State Oil Company of the Republic of Azerbaijan (SOCAR), Rovnag Abdullayev, Azerbaijan was the first country to use onshore and offshore oil production and the first to use tankers for oil transportation (Abdullayev, 2009). Indeed, the world's first oil tanker was designed and built in Baku by Ludwig Nobel in the late-nineteenth century (Yergin, 2011, p. 50).

At the time, the country's oil production saw an influx of foreign investors and was controlled by oil barons (Taghiyev, Naghiyev, Mukhtarov, Hajinsky), such as the Nobel brothers and the Rothschilds. The Nobel brothers' company was the world's largest oil company. In 1901, Azerbaijan's oil production (80.5 million barrels) exceeded American output (63.7 million barrels) and comprised more than half of the world's oil production. The capital, Baku, was booming because of the oil revenues. After Azerbaijan was annexed by the Soviet Union in 1920, oil production boom in Baku continued. Pre-WWII, the Azerbaijan Soviet Republic produced 75 per cent of the overall Soviet oil output. After the end of the Soviet Union in 1991, Azerbaijan opened its resource-based economy and embraced investments and guidance from the West, and especially from the United States (I. Mammadov, 2010).

Data on Azerbaijan's oil reserves have varied over the years, depending on the source of information. The Energy Information Administration (EIA) of the U.S. Department of Energy initially assessed the overall Caspian oil reserves at 233 billion barrels but later drastically reduced the estimates to 17–33 billion barrels. In 2002, Chairman of the Italian oil company ENI, Gian Maria Gros-Pietro, announced that the whole Caspian region possessed only 7.8 billion barrels of recoverable oil, with one third of it located in Azerbaijan (Blanchard, 2005, p. 206). Energy analysts have noted that estimates of oil countries' reserves often become inflated in the literature, and modern technology is not advanced enough to measure them with precision (Shaffer, 2009a). For example, the Kashagan field in Kazakhstan was initially rated as a supergiant, but the actual drilling uncovered dry wells, and the estimates were lowered (Blanchard, 2005, p. 14).

For the last two decades, despite conflicting evaluations of the country's oil reserves, Western energy companies and governments came to Azerbaijan for new contracts, attracted by the country's political stability and an improved business climate. Since then, oil production has steadily grown with foreign capital and technology, and oil revenues have constituted half of the country's GDP. The U.S. Energy Information Administration (EIA) reported that in over ten years, Azerbaijan's oil output grew almost ten-fold from 150,000 barrels per day (in 1997) to 1.04 million barrels per day (in 2009). Between 1992 and 2004, oil production was fluctuating between 150,000 and 350,000 barrels per day. After 2005, however, it soared to 850,000 barrels per day by 2007, thanks to Western investments ('News in Brief', 2009). The ACG oil field alone was estimated to produce 1,000,000 barrels per day at its peak (Blanchard, 2005, p. 7).

Currently, oil revenues are 70 per cent of all Azerbaijan's exports and 50 per cent of state revenues. In 2015, Azerbaijan produces 881,000 barrels per day, and experts predict that the oil production will peak (i.e. will reach its maximum and then start declining) at 894,000 barrels per day in 2017, because one of the main oil fields, Azeri-Chirag-Guneshli (ACG), is experiencing rapid depletion. In 2024, Azerbaijan is predicted to produce 762,000 barrels per day ('Azerbaijan Petrochemicals Report – Q2 2015', 2015, p. 23).

Azerbaijan is also a major natural gas producer. The ACG and the Shah Deniz fields currently provide the bulk of Azerbaijan's natural gas production. The Shah Deniz II project, which is currently underway, will increase the country's gas production in 2018, when the project is scheduled to start operations. The ACG field, as well as the Absheron gas field, will provide 12–30 billion cubic meters of natural gas at the peak of production in 2021–2022 ('Azerbaijan Petrochemicals Report – Q2 2015', 2015, p. 23).

Azerbaijan's current challenges related to oil and gas exports are somewhat similar to those the country experienced in the nineteenth century: the country's export routes are limited to the Caspian Sea region; the Caspian is an area of strong competition among great powers; and the region is plagued by ethnic and secessionist conflicts (Shaffer, 2009b, pp. 68–69). On the other hand, new technology in transportation and drilling have provided new opportunities for Azerbaijan to be not only a producer of oil and natural gas but also a transit state for resources from Turkmenistan and Kazakhstan (Pashayev, 2009, p. 110).

In addition to ethnic disputes on land, such as Nagorno-Karabakh, oil and gas development in the Caspian is complicated by an ongoing dispute among the littoral states (Azerbaijan, Kazakhstan, Turkmenistan, Iran, and Russia) over the status of the Caspian Sea. They disagree whether the 750-mile-long and 200-mile-wide Caspian Sea is a sea or a lake. This legal dispute is difficult to resolve because the science does not give us a definite answer. Scientists have concluded that the

Caspian Sea, as a water body, has qualities both of a sea and a lake. This distinction between a sea and a lake is important to establish, because the inter-state divisions of seas and lakes are governed by two different international groups of laws. Russia has considered the body of water as a sea, and Azerbaijan – as a lake. If it is a lake, it needs to be divided into national sectors. Both Azerbaijan and Kazakhstan benefit from the division into national sectors, because it gives them large offshore hydrocarbon reserves. The other three littoral states benefit from the sea arrangement (LeVine, 2001).

Among the littoral states, Turkmenistan eventually accepted that the Caspian Sea is a lake but disagreed with Azerbaijan on where the median line dividing the body of water into national sectors should be drawn. Russia offered an alternative solution: to divide only the seabed, while treating the water surface as an area of common use. Iran made several propositions, ranging from a condominium regime to joint exploration of resources. These disagreements among states created uncertainties for future oil and gas exploration in the region, because international oil companies did not want their exploration rights taken away from them (I. Mammadov, 2010). In 2003, Russia, Azerbaijan, and Kazakhstan reached an agreement to divide the oil-rich northern part of the Caspian Sea into three parts: Kazakhstan received 27 per cent, Russia – 19 per cent, and Azerbaijan – 18 per cent. Based on these borders, the U.S. Department of State estimated that Azerbaijan's proven oil reserves were 7–13 billion barrels (Blanchard, 2005, pp. 206–208).

As oil production was increasing, a crucial issue to solve was the transportation of energy resources from Azerbaijan to the West. Azerbaijan has long worked on developing export routes for its oil, dubbed as the 'policy of multiple pipelines': the Azerbaijani government explored several potential routes in the early 1990s in order to deliver oil from Azerbaijan to the Western European market. The first route was from Azerbaijan to the north, to Russia. This option was not supported by the Azerbaijani government and by Europe because the main idea of the 'policy of multiple pipelines' was to bypass Russia. The second option was to go south, to the northern border of Iran, where oil would be delivered to Iranian refineries and swapped with Iranian oil in the southern part of Iran. This option was also unfeasible, as both the United States and Azerbaijan did not want to depend on Iran's will to fulfill future Iranian obligations to swap the oil. In addition to that, oil would have to be delivered to world markets through the already overcrowded Persian Gulf.

The third option was the Baku–Tbilisi–Ceyhan route, actively supported by the United States, as well as by Europe, even though experts estimated that it was the most expensive and technologically challenging of all the three options (Yergin, 2011, p. 59). Eventually, this route was selected, mostly on geopolitical grounds. On October 29, 1998, the governments of Georgia, Turkey, Kazakhstan, and Uzbekistan signed the Ankara Declaration, which defined the route of the Baku–Tbilisi–Ceyhan (BTC) pipeline. Parallel to BTC, there was supposed to be a gas pipeline (the South Caucasus pipeline) designed to bring natural gas from the Shah Deniz field to Ceyhan (Yergin, 2011, p. 63).

Analysts warned, however, that Azerbaijan's current reserves might not be sufficient to fill all the planned pipelines (BTC, Nabucco): Azerbaijan's oil production was almost at its peak and building new pipelines would have a substantial economic risk. BTC, for example, was designed to transport oil not only from Azerbaijan but also from Kazakhstan, in order to ensure the promised volume of exports. Kazakhstan, however, might divert its resources to other projects in the future, including oil swaps with Iran (Sabonis-Helf, 2009).

To increase Azerbaijan's energy trade diversification, Foreign Minister of Azerbaijan, Elmar Mammadyarov, stressed in 2009 that Azerbaijan must expand the number of its consumers of oil and gas, in order to strengthen the country's sovereignty, simultaneously making sure that the projects are profitable. He noted that it was much easier to diversify the oil trade, as opposed to the natural gas trade, because oil was easily transportable. Natural gas can be transported globally as liquefied natural gas (LNG), but it requires an extensive new infrastructure, such as pipelines, LNG plants, and tankers for LNG transportation (Mammadyarov, 2009a).

More recently, Azerbaijan has continued its expansion of the pipeline system. In addition to BTC, starting in 2019, natural gas from Azerbaijan will be transported via the Trans-Anatolian Pipeline (TANAP) from the Shah Deniz II project, in the capacity of 10 billion cubic meters per year. The pipeline could potentially transport even more, about 20 billion cubic meters per year, if Gazprom or another producer decides to use the additional capacity for their needs, paying Azerbaijan for the transportation ('Emerging Europe Oil and Gas Insight – May 2015', 2015, p. 1).

After long discussions and negotiations, the construction of another pipeline, the 1,850-km Trans-Anatolian Pipeline (TANAP), started in March 2015 and is supposed to be completed in 2018. The pipeline, as part of the Southern Gas Corridor, will connect with TAP and deliver natural gas produced by Azerbaijan's Shah Deniz II project to Turkey, Albania, Italy, and Greece. Potentially, the web of pipelines will connect Europe with Turkmenistan and Iran, depending on future political relations with those countries. The TANAP has been designed as a step to reduce Europe's imports of Russian natural gas. The TANAP will deliver about 31 billion cubic meters of gas in 2026, which is about ten per cent of the current European imports of 275 billion cubic meters per year ('Eastern Europe Politics: Construction Begins of Trans-Anatolian Pipeline', 2015).

Azerbaijan's supplies of natural gas currently compete with Russia's potential imports to Turkey, and the competition with Russia could bring natural gas prices in Turkey and Azerbaijan down. This will benefit the Turkish and European economies but will reduce export revenues for the Azerbaijani government ('Eastern Europe Politics: Construction Begins of Trans-Anatolian Pipeline', 2015). A similar effect on gas prices might happen, if Turkmenistan and Iran further increase their exports to Europe, bringing prices down even more. Azerbaijan has competed with Russia not only in natural gas pipelines, but also in oil. The Odessa–Brody pipeline, for example, brings Caspian oil to Slovakia, Czech Republic, and Hungary, reducing Russia's share of the European energy market (Zygar & Grib, 2009). Azerbaijan's position is especially favorable in the recent turmoil between Russia and the West, while Azerbaijan has maintained a favorable economic climate for foreign investors.

Investment climate

Oil and gas reserves alone are not sufficient to determine the opportunities for foreign companies. To invest and operate in a host country, the companies consider a set of conditions that is usually defined as an investment climate, i.e. regulations enhancing or constraining business activity. This bundle of conditions shows how easy it is for foreign companies to do business in a certain country.

Azerbaijan's domestic political stability in itself already attracted foreign investors, and the country implemented other special measures to attract foreign companies. Some of those measures and initiatives aimed at promoting transparency and decreasing corruption. The government put emphasis on the protection of foreign investments from changes in the economic and political climate, signing the Law on Protection of Foreign Investments and joining the Extractive Industries Transparency Initiative (EITI) ('Doing Business In Azerbaijan: A Country Commercial Guide for U.S. Companies', 2008).

The Law on Protection of Foreign Investments, passed by the government of Azerbaijan in 1992, shortly before the 1995 Law on Investment Activity, stipulates that if the government adopts new regulations unfavorable for foreign investors, the investors are protected by a 10-year grandfather clause (with the exception of tax policies) from any nationalization and property requisitions (except for emergency situations). If such requisitions have to happen, the foreign companies are entitled to fair compensation ('Doing Business In Azerbaijan: A Country Commercial Guide for U.S. Companies', 2008, p. 42). The Law on Protection of Foreign Investments makes sure that foreign investors have the same rights as domestic companies do. Foreign investors are guaranteed that their profits could be freely transferred to their home countries (Sallis, 2003).

The EITI, which Azerbaijan joined in 2007, was first started by the United Kingdom in 2002 at the World Summit on Sustainable Development, followed by two other conferences in London in 2003 and 2005. The initiative's goal is to ensure that the public has effective oversight of the extractive enterprises and natural resources (not only oil and natural gas, but also all kinds of commodities – timber, metals, and others). Major corporations, such as ConocoPhillips, BP, Chevron, Royal Dutch Shell, and others have joined the initiative over the years. Professional associations of extractive industries, such as the International Council on Mining and Metals, as well as non-profit organizations such as Human Rights Watch and Global Witness, also supported the initiative (Lee, 2006).

These laws and initiatives help build foreign investors' trust, and the investments continued to flow to Azerbaijan, from $330 m in 1995 to $2 bn in 2003 and 2004 ('Doing Business In Azerbaijan: A Country Commercial Guide for U.S. Companies', 2008, p. 45). International energy companies and the Ministry of Energy and Industry of Azerbaijan signed production sharing agreements (PSAs) for oil and natural gas exploration. The agreements were granted the status of national laws, in order to give the investors even more assurances.

PSAs were signed in the context of bilateral trade agreements that the government of Azerbaijan was negotiating with multiple states. In 2000, an agreement was signed and ratified with the United States, in order to encourage and protect the countries' mutual investments. As of 2015, a bilateral trade agreement is being discussed with the European Union: such an agreement would be extremely important for Azerbaijan, as the 42 per cent of Azerbaijan's overall export volume goes to the European Union. Also, 28 per cent of Azerbaijan's overall imports come from the EU. The agreement will expand the scope of the previous trade protocol – the EU–Azerbaijan Partnership and Cooperation Agreement (PCA) of 1999. Unlike the PCA, the new agreement would give Azerbaijan more preferential conditions in its trade with the European Union and create a 'more rule-based business environment' ('New Azerbaijan–EU Agreement to Help Expand Trade', 2015).

Azerbaijan's government also simplified business registration procedures, reducing the red tape required to open a new business. As a result, Azerbaijan's score in 'Doing Business' rankings has been similar to Turkey's and South European countries' and even better than Russia's. One of the parameters – the time required to open a new business – improved in Azerbaijan from 122 days to 16 days, and the cost to open a new business fell by 77 per cent ('Doing Business In Azerbaijan: A Country Commercial Guide for U.S. Companies', 2008, pp. 40–41).

Simplified procedures made it easy to finance large-scale projects, in which governments, energy companies, and international institutions combined their resources. Financing the $3.6-billion BTC pipeline, for example, involved multiple investors who reached a compromise on sharing costs and profits. BP owned 30 per cent of equity, SOCAR – 25 per cent, UNOCAL – 8.9 per cent, Statoil – 8.7 per cent, TPAD – 6.5 per cent, and several others owned a smaller percentage. The whole amount of equity was $1.1 bn, while the amount of debt capital was $2.5 bn (Razavi, 2007, p. 278).

The Azerbaijani government has worked closely with international institutions, such as the International and the European Banks for Reconstruction and Development (IBRD and EBRD), the International Monetary Fund (IMF), and the Asian Development Bank (ADB). For the last several years, Azerbaijan has also been working on joining the World Trade Organization (WTO) ('Doing Business In Azerbaijan: A Country Commercial Guide for U.S. Companies', 2008, p. 2). Under the guidance of these international institutions, Azerbaijan has invested in its domestic economy and infrastructure (education, health care, foreign debt repayment) (Sadigov, 2009, p. 140). Public spending has contributed to a higher rate of employment and better education, which are favorable factors for foreign companies who come to Azerbaijan (Maharramov, 2010, p. 38).

One of the international institutions that monitor the investment climate in most countries of the world is the World Bank. It does so by publishing extensive reports, as well as ranking countries in terms of the easiness of doing business. In

2008, the World Bank's 'Doing Business' report named Azerbaijan one of the top ten *reformers* in Eastern Europe and Central Asia. Azerbaijan's place jumped from the 97th in 2008 to the 33rd in 2009, which was a large improvement in the easiness of doing business in the country ('Doing Business 2009', 2008, p. 18). In 2014, Azerbaijan held the 70th place among 189 countries in the world in the 'Doing Business' ranking. The improvement in the ranking resulted from 18 reforms in the country's regulation and legislation. Other improvements, such as electronic registration of businesses, placed Azerbaijan in the top ten countries with the best conditions for starting a business. The Azerbaijani government also implemented a 'one window' policy for international trade applications (Khalilova, 2013).

In the United States, the U.S. Department of State also produces reports about doing business in various countries, including Azerbaijan. The U.S. government has monitored Azerbaijan's economy since the early 1990s. In 1994, a resource guide for U.S. businesses described Azerbaijan as a country with substantial energy resources, moving 'cautiously' towards a market economy, but still struggling with an underdeveloped infrastructure, a lack of hard currency reserves, and profit expatriation barriers ('Doing Business in the New Independent States – A Resource Guide', 1994). In 2008, a U.S. government report concluded that Azerbaijan was 'a challenging market in which to do business' ('Doing Business In Azerbaijan: A Country Commercial Guide for U.S. Companies', 2008, p. 41). In 2014, the U.S. Department of State stated that 'successive administrations have supported U.S. private investment in Azerbaijan's energy sector as a means of increasing the diversity of world energy suppliers' (Nichol, 2014). The U.S. government has been mainly concerned about the Azerbaijan–Armenia dispute over Nagorno-Karabakh, as well as human rights issues in the region (Nichol, 2014).

The first major contract between Azerbaijan and foreign investors was signed on September 20, 1994, in response to Azerbaijan's newly acquired political stability. The contract was so important that it was called the 'Contract of the Century', which stipulated the terms and conditions of energy production and trade between SOCAR and international oil companies. This deal of the century, aimed at exploration of the ACG oil and gas field, was the first not only in Azerbaijan, but also in the whole Caspian region to bring together a diverse group of governments and companies (Yergin, 2011, p. 54).

The signing of the contract involved an impressive list of dignitaries and signatories. The high-level signatories were Tim Eggar, UK Energy Minister; Heydar Aliyev, President of Azerbaijan; Bill White, US Energy Deputy Secretary; Usam Jafari, Islamic Bank of Development; Stanislav Pugach, Russian Ministry of Fuel and Energy; Thomas Young, UK Ambassador; Richard Kauzlerich, US Ambassador; Eldar Namazov, Presidential Advisor; and Hasan Hasanov, Foreign Minister of Azerbaijan. Executives of major energy companies also participated in negotiations and signing (Sagheb & Javadi, 1994).

The contract stipulated the stakes and profit divisions among the participated parties. SOCAR was entitled to 80 per cent of all profits from the ACG oil production. The remaining 20 per cent of the profits was to be divided among multiple participants: SOCAR (Azerbaijan) – 20 per cent; British Petroleum (UK) – 17 per cent; Amoco (USA) – 17 per cent; LUKoil (Russia) – 10 per cent; Pennzoil (USA) – 9.8 per cent; Unocal (USA) – 9.5 per cent; Statoil (Norway) – 8.6 per cent; McDermott International (USA) – 25 per cent; Ramco (Scotland) – 2 per cent; Turkish State Oil Company (Turkey) – 1.8 per cent; Delta-Nimir (Saudi Arabia) – 1.7 per cent. It was expected that Azerbaijan would receive $81 bn over 30 years from this deal (Sagheb & Javadi, 1994).

The Contract of the Century was followed by other international contracts between Azerbaijan and major energy companies. In September 1994, a number of oil companies – BP (UK), Chevron (USA), Devon Energy (USA), Amerada Hess (USA), ExxonMobil (USA), StatoilHydro (Norway), Türkiye Petrolleri Anonim Ortaklığı – TPAO (Turkey), SOCAR (Azerbaijan), Inpex (Japan), and Itochu (Japan) – formed a consortium called the Azerbaijan International Operating Company (AIOC), to develop three oil and gas fields: the Azeri, Chirag, and Gunashli (ACG) fields ('Shah Deniz Signed: Third Major Contract in Caspian', 1996).

Another large contract was signed on September 20, 1994 for the development of the Azeri, Chirag, and Gunashli (ACG) fields. The project participants and operators were BP (which held a 34 per cent interest in the project), Chevron (10 per cent), State Oil Company of the Azerbaijani Republic (SOCAR) (10 per cent), Inpex Corp. (10 per cent), Statoil (8.6 per cent), ExxonMobil (8 per cent), TPAO (6.8 per cent), Devon Energy (5.6 per cent), Amerada Hess (2.7 per cent), and ITOCHU Oil (3.9 per cent) ('Russia', 2010).

A year later, in 1995, development started in another substantial oil field – the Karabakh Prospect. The operator of this project was the Caspian International Petroleum Company (CIPCO), which included such partners as LUKAgip (45 per cent of the stake in the project), Pennzoil Caspian Development Corporation (30 per cent), LUKoil International Ltd. (12.5 per cent), a SOCAR commercial affiliate (7.5 per cent), and Agip Azerbaijan (5 per cent) ('Shah Deniz Signed: Third Major Contract in Caspian', 1996).

In 1996, the Azerbaijani government started to develop the Shah Deniz oil field. The participating companies included British Petroleum (25.5 per cent of the stake in the project), Statoil (25.5 per cent), LUKoil (10 per cent), ELF (10 per cent), the National Oil Company of Iran (10 per cent), Turkish Petroleum (10 per cent), and SOCAR (9 per cent). As it was during the signing of the Contract of the Century, the signatories of the Shah Deniz deal also included many prominent government and business representatives, who had a common interest of securing a stable supply of energy resources from Azerbaijan. The list of attendees at the contract signing is indicative of the importance that Azerbaijan had on the international energy market: Guy Arlette, Deputy Minister of Energy and Industry (France); Vladimir Kostyushin, First Deputy Minister of Energy (Russia); Heydar Aliyev, President (Azerbaijan); Tim Eggar, Minister of Energy (UK); Seyid Ali Akbar Hashimi, President of Iranian Oil Company; Thomas Young, Ambassador (UK); Ömür Orhun, Ambassador (Turkey); Sidgi Süncär, President of Turkish Petroleum Company; Ravin Maganov, VP of LUKoil (Russia); Philippe Jaffre,

President of ELF Aquitane (France); Bayron Grodini, VP of BP (UK); Natig Aliyev, President of SOCAR (Azerbaijan); Jon Nic Vold, Executive VP of Statoil (Norway); Anwar Saifullah Khan, Minister of Oil & Natural Resources (Pakistan); Guy Arlette, Deputy Minister of Energy and Industry (France); Lucian Motiu, State Secretary responsible for Oil and Gas, Ministry of Industry (Romania); Baltabay Kuandykov, President of State Oil Company (Kazakhstan); K. Saldastanishvili, Minister for Economic Relations (Georgia); and Jens Stoltenberg, Minister of Industry and Energy (Norway) ('Shah Deniz Signed: Third Major Contract in Caspian', 1996).

British Petroleum (BP) – a participant of most of those projects – has been one of the most active in the region. The company started its operations in Azerbaijan in the 1990s and quickly became the leader in production and transportation of energy resources in the country. BP's long-term interests in the region include the ACG oil field (in the Caspian Sea) and the Shah Deniz gas field. In 2009, BP Vice President described Azerbaijan as a country with low political risk. He added, though, that oil production there had technological challenges (Khalilov, 2009). According to BP-Azerbaijan's Vice-President for Environment, Greg Mattson, BP's 'holy trinity' (energy, technology, and policy) brought new technology to Azerbaijan, ensured stable oil revenues for the government, and promoted sustainable development. He positively evaluated Azerbaijan's business environment that granted a status of law to PSAs and ensured the stability of contracts (Mattson, 2009).

As of 2015, BP continues participating in major projects and joining new ones. The company joined the Trans-Anatolian Natural Gas Pipeline (TANAP) with a 12 per cent stake on April 12, 2015. The main stakeholder in the project is SOCAR (50 per cent), followed by Botas (Turkey) – 30 per cent ('BP Becomes Stakeholder in Azeri-led Regional Gas Project', 2015).

For comparison, BP has not had such smooth operations in Russia. Earlier in 2008, Robert Dudley, the CEO of the TNK-BP joint venture had to leave Russia after he had been accused by the Russian company AAR (owned by Mikhail Fridman, Len Blavatnik, and Viktor Vekselberg) of acting in the interest of the British shareholders. In 2010, BP's CEO at the time, Tony Hayward, took a position of nonexecutive director of TNK-BP, while Dudley became the CEO of BP. Russia's Deputy Prime Minister at the time, Igor Sechin, commented on Hayward's appointment: 'We have developed good and warm relations during [Dudley's] time in Russia. He knows the Russian market well'. Nevertheless, a couple of years later, in 2012, BP had to sell its stake in TNK-BP, which was bought out by the Russian side of the joint venture (Razumovskaya, 2010).

Russian companies have participated in some of the Caspian energy projects. Russia's Blue Stream gas pipeline on the bottom of the Black Sea delivers Azerbaijan's Shah Deniz natural gas from Turkey to Bulgaria, Hungary, and Romania ('Russia Scraps South Stream Natural Gas Pipeline Project', 2014). The Russian company LUKoil, led by Baku-born Vagit Alekperov, has held a partial ownership of upstream projects in Azerbaijan. However, LUKoil decided not to join the BTC project and later, in 2002, sold LUKoil's share in the ACG field (Nanay, 2009). LUKoil is still involved in the Shah Deniz projects, though. Currently, as of 2015, LUKoil owns 10 per cent of the Shah Deniz contract, signed on June 4, 1996, together with BP (28.8 per cent), TPAO (19 per cent), Statoil (15.5 per cent), AzSD (10 per cent), NICO (10 per cent), and SGC Upstream (6.7 per cent). Natural gas from the project is delivered to Turkey and the European Union ('New Well Within Largest Gas Project Drilled in Azerbaijan', 2015).

In sum, Azerbaijan created a favorable investment climate for foreign companies. In 2010–2014, foreign direct investments (mostly in the energy sector) were 8 per cent of the country's GDP, which was significantly higher than the percentage in other post-Soviet nations ('Azerbaijan: Country Fact Sheet', 2015).

The economic and environmental effects of oil and gas projects

The international and domestic projects for oil and gas exploration have brought significant oil revenues to Azerbaijan. Azerbaijan increased its state budget from $1.2 billion in 2003 to ten times that amount ($11 billion) seven years later (Guliyev, 2013). Similar to other oil-rich nations, such as Russia and Norway, Azerbaijan has accumulated a part of oil and gas revenues in a special fund – the State Oil Fund (SOFAZ). It was initially started in 1999, when the fund held only about $270 million (Ismayilov, 2014). As of 2012, the fund accumulated more than $30 billion. Oil revenue transfers from SOFAZ to the country's state budget constituted about 44 per cent in 2010 and 50–60 per cent in 2012 (Guliyev, 2013). In mid-2014, the fund had already $37 billion (Ismayilov, 2014). In 2014, the share of SOFAZ money transferred to public spending reached 80 per cent of the government's financing for public projects ('Azerbaijan: Vested Interests Will Resist Change', 2014). Using SOFAZ, the government substantially invests in social and economic infrastructures, such as housing, education, and transportation. As a result, the Word Bank and the International Monetary Fund regularly evaluate Azerbaijan as one of the fastest growing economies in the Caspian region (I. Mammadov, 2010).

International organizations and foreign companies have praised Azerbaijan's prudent management of oil revenues. Using the low-risk strategies of saving the oil revenues and using SOFAZ for domestic projects, Azerbaijan stayed resilient during the global recession that hit most countries in 2008 (Khalilov, 2009). In 2009, Azerbaijan's GDP even increased by about 9 per cent, partly because oil and natural gas production (mainly from the Azeri–Chirag–Guneshli (ACG) fields) jumped 13 per cent in that year. Other post-Soviet and Eastern European countries took much longer to recover ('Azerbaijan: Country Outlook', 2010).

Experts agree that several factors have contributed to Azerbaijan's economic resiliency. First, the Azeri currency, manat, has been pegged to the U.S. dollar since 1994, stabilizing the Azeri economy (although the peg has also been making Azeri oil production more expensive because of the rising value of the dollar) ('Azerbaijan', 2006). The country has used a *soft peg*, or a *crawling peg*, policy, which means that the Central Bank of Azerbaijan has had an *exchange rate target* for the U.S. dollar and the Euro, allowing some controlled variation in the exchange rate. As a result of this policy, the manat did not experience as much damage as other currencies during the recession of 2008 (Dabrowski, 2013).

Second, the Azeri government adopted conservative fiscal and investment policies, avoiding profitable but volatile financial instruments (Khalilov, 2009). In 2009, Azerbaijan was the only state in the region with the non-oil-sector economic growth of 4.8 per cent (and 9 per cent in the energy sector alone) (Ibadoglu, 2011, p. 4). The growth was induced by government investments in infrastructure and other public projects. The fiscal consolidation reform and subsidies have boosted the economy, especially in transportation, agriculture, and other sectors. However, more money available to consumers has led to a higher level of inflation ('Azerbaijan: Vested Interests Will Resist Change', 2014). As of 2014, the average rate of inflation in the country was 3.7 per cent, which is still lower than in other post-Soviet nations ('Azerbaijan: Country Fact Sheet', 2015). In spending oil revenues, the government has kept the balance between short-term spending and long-term country development goals, in order to keep the living standards growing and to maintain social stability in the country (Kendall-Taylor, 2010, p. 16).

Azerbaijan's economy highly depends on oil and natural gas exports. In 2013, petroleum products comprised 92.8 per cent of all exports from Azerbaijan. The next largest industry, food products and animals, was only 3.7 per cent of the overall export volume ('Azerbaijan: Country Fact Sheet', 2015). Experts in the region pointed to the fact that the non-oil sectors (such as steel and metal-rolling) have been the so-called 'necroeconomic enterprises', which means that they have operated at the lowest possible capacity or stayed idle (Papava, 2011, pp. 22–23).

The economy's dependence on oil and natural gas has been challenging in terms of balancing economic growth with environmental protection. The country's capital, Baku, has been heavily polluted since oil production started in the nineteenth century. According to Azerbaijan's Minister of Ecology, Huseynguly Bagirov, the Caspian Sea bed currently contains from 3 to 10 meters of oxidized oil products, accumulated during the decades of oil production. Talking about the BTC pipeline, Bagirov indicated that the pipeline was not an ecological 'disaster', but British Petroleum and other companies should do more to protect the environment. There have been attempts in Azerbaijan to form a government-led environmental coalition of NGOs and businesses, in order to make oil companies fully comply with the environmental law and pay multi-million dollar environmental penalties if they break the environmental law (Bagirov, 2009). In compliance with environmental regulations, oil companies have come up with various initiatives to make their business 'greener'.

In sum, the government of Azerbaijan has prudently utilized Azerbaijan's resources and the geographic closeness to Western Europe. President Heydar Aliyev's decision in 1994 to turn to the West shortly after the country's independence, brought very positive results. He invited foreign investors to participate in the country's energy sector, making sure that the investor's rights are well protected by laws and production sharing agreements. This strategy has helped turn Azerbaijan, which had been a natural gas exporter before the independence, into a natural gas importer. The recently developed Shah-Deniz gas field and others provide natural gas that is sold to Turkey, Russia, Georgia, and the European Union, contributing to European energy security. The Azerbaijani government has also repeatedly emphasized that the energy trade contributes to the country's policy priorities beyond the energy sector, such as infrastructure development and military spending, especially in light of the conflict with Armenia over Nagorno-Karabakh (Abdullayev, 2009).

Geopolitical rivalries in the Caspian

The investment climate in Azerbaijan has been favorable for foreign companies, which were provided strong guarantees by the Azerbaijani government, and this partly explains why Azerbaijan has been so successful in cooperating with the West and especially with the United States. However, when it comes down to the specific details of energy deals, such as a pipeline's route selection, then the economic indicators, such as 'Doing Business' rankings, matter less. The choice of trade routes and partners is also shaped by geopolitical factors, especially by the U.S.–Russia rivalry in the region, as well as by the EU's desire to bypass Russia and to reduce the reliance on Russian natural gas. Ethnic conflicts also affect decision-making: one of such conflicts has been Nagorno-Karabakh.

Nagorno-Karabakh

A protracted conflict between Armenia and Azerbaijan over the Nagorno-Karabakh region, which comprises 20 per cent of Azerbaijan's territory, started in 1988 (Tariverdiyeva, 2015). Since then, the restoration of Azerbaijan's territorial integrity became the main security priority for Azerbaijan and the central point of the country's foreign policy (Alieva, 2011, p. 198).

The roots of the conflict go back to a century ago, when the Soviet leader Joseph Stalin, established artificial borders between Soviet republics, dividing nations, thus ensuring their compliance. In 1924, the Soviet central government decreed the predominantly Armenian-ethnic Nagorno-Karabakh an autonomous entity within the Soviet Republic of Azerbaijan. It was done despite the fact that the share of the Armenian populations in Nagorno-Karabakh was as high as 94 per cent. The Armenians were a majority in local legislative bodies, controlling the political power of the region. During the bulk of the Soviet period it did not create much problem, however, and the Armenians and the Azerbaijanis peacefully coexisted in Nagorno-Karabakh ('Nagorno-Karabakh', 2009).

Everything changed in 1988. In February of that year, Nagorno-Karabakh's legislative body (dominated by Armenian deputies) made a decision to unite the Nagorno-Karabakh region with Armenia. There were several official reasons for this decision. First, the population of the region was predominantly Armenian.

Second, the Armenian population in Nagorno-Karabakh had complained about discrimination against them by Heydar Aliyev's Soviet government back in the 1970s and 1980s. This decision caused violence between the Armenians and the Azerbaijanis and a massacre in the city of Sumgait in February 1988. After the massacre, Azerbaijan announced that Nagorno-Karabakh was under occupation by the Armenian army ('Nagorno-Karabakh', 2009).

The conflict escalated over the next several years. In an attempt to return Nagorno-Karabakh to Azerbaijan, the Azerbaijani government established an economic blockade of the region. However, the blockade did not change the situation – Nagorno-Karabakh stayed under Armenian control. When the USSR dissolved in 1991, the conflict turned into a full-scale war, in which one million Azerbaijani people became refugees or internally displaced people. Heavy fighting happened in 1990–1994 at key strategic sites, such as the Lachin corridor – the town and its surrounding area connecting the Nagorno-Karabakh. To Azerbaijan's dismay, Armenia received assistance from Russia and the United States, thanks to close ties with both countries. From 1993 to 1995, Russia provided Armenia with arms and continuously supplied the country with natural gas ('Nagorno-Karabakh', 2009).

The attempts to reconcile the two sides started as early as in 1989 by the last Soviet leader Mikhail Gorbachev but have repeatedly failed. The United States and Russia have both tried to mediate the conflict through economic and diplomatic methods, but also failed, as their efforts were not coordinated. The international community, such as the Organization for Security and Co-operation in Europe (OSCE), has also long participated in Azerbaijan–Armenia negotiations (Tariverdiyeva, 2015).

The latest attempt happened on April 30, 2015, when the so-called OSCE Minsk Group (co-chaired by France, the Russian Federation, and the United States) had individual meetings with the Foreign Minister of Azerbaijan, Elmar Mammadyarov and Foreign Minister of Armenia, Edward Nalbandian. Before that, both ministers met separately with President of France, Francois Hollande. After the meetings, the Minsk Group issued the following statement: 'None of our three countries, nor any other country, recognizes Nagorno-Karabakh as an independent and sovereign state' (Tariverdiyeva, 2015).

At present, Nagorno-Karabakh remains a frozen conflict that is increasingly difficult to solve. Two generations of Azerbaijani and Armenians have not talked to each other face to face, and international attempts have not brought desired results. This conflict has an implication for the energy trade that Azerbaijan does with other states: Armenia is not included in pipeline routes and energy projects.

Diplomatic maneuvering among great powers

Early on, the government of Azerbaijan understood the need to remain independent of the surrounding states (Russia, Israel, Iran, and Turkey), which was not a simple task considering that Azerbaijan is a small state of only 9.5

million people. Small states usually choose one of several priorities in their foreign policies: an active participation in international organizations; self-reliance; alliance-building; and strategic diplomatic maneuvering (Mehdiyeva, 2011, p. 18). Azerbaijan has taken the fourth path – especially in energy policy and military strategies. The government has always realized that in order to advance energy trade and sell Azerbaijan's energy resources, the country needs to keep productive relations with all their neighbors, at the same time keeping the conflict in Nagorno-Karabakh from escalating.

Energy security has long been at the center of Azerbaijan's geopolitical priorities and national interests. Elin Suleymanov, then Consul General in Los Angeles, California, and now the Ambassador of Azerbaijan, said it succinctly: 'Azerbaijan is a *pragmatic* state that has followed Heydar Aliyev's strategic vision and promoted the country's national interest' (Suleymanov, 2009). These strategic interests were especially obvious during the negotiations of the $4-billion, 1,099-mile BTC pipeline route selection.

One of the geopolitical priorities for Azerbaijan has been to keep strong relations with the United States. The interest between the two countries is mutual: Azerbaijan needs a strong partner, and the United States needs an alternative source of oil and natural gas in the region, as well as a military ally. Already in 1997, the U.S. Department of State, in its report entitled 'Energy Development in the Caspian region', summarizes such priorities as the expansion of the world's energy supply and demand; the sovereignty and independence of the Caspian Basin countries; and the need to isolate Iran (Nassibli, 2004, p. 167). Since then, the priorities have remained almost the same. The BTC pipeline, built in 2006, enjoyed the 'enormous U.S. government political backing' that encouraged major energy companies 'to get energy to market through private pipelines bypassing Russia' (Nanay, 2009, p. 117).

The desire to bypass Russia stemmed partly from the criticism by American experts of Russia's policies in the Caspian. Edward Lucas, in his *New Cold War*, stressed that Azerbaijan had built the BTC pipeline to avoid the Kremlin's 'imperial embrace' and control. As previously with BTC, the United States strongly supported the planned Nabucco pipeline from Turkey to Austria. A senior U.S. adviser for Caspian Basin Energy Diplomacy, Steven Mann stated that the choice of pipeline routes and trade partners was a matter of sovereignty for Caspian states (Mann, 2003, p. 153).

The American government has supported Azerbaijan's independence from the influence by other Caspian states and, at the same time, Azerbaijan's integration into the regional energy web. The main purpose was to 'keep Russia and Iran in check', by providing Azerbaijan with a prominent role in energy transportation (BTC and other pipelines). Turkey, another reliable NATO partner, was encouraged to take on the role of the transportation hub for the regional oil and gas routes (Nassibli, 2004). When BTC was in construction, experts were concerned that Russia would see 'Western influence in the Near Abroad as an attempt to further undermine Russia and retard the restoration of its Great Power status' (Yergin, 2011, p. 47). However, Russia did not block the BTC pipeline.

Iran has been another country that has been 'kept in check', because of its confrontations with the United States, both on the issues of energy supplies (oil and natural gas trade and nuclear energy development) and on security concerns (potential development of nuclear weapons). As was mentioned earlier, the Iranian route was one of potential options for Azerbaijan's oil and natural gas exports. However, the U.S. government opposed the Iranian route, because depending on Iran for oil swaps would make Azerbaijan and the West too vulnerable and would defeat the purpose of flexibility and energy independence.

For years, Iran had tried to enter the Caspian gas market. Mahmood Khaghani, Chief of the Caspian department in the Iranian Oil Ministry, stressed that new pipelines, such as Nabucco, would be more cost efficient and economically justifiable if the Iranian natural gas was allowed in. He also argued that BTC was not economically feasible and was built on political grounds to bypass Russia. The disagreements between Iran and the West on everything from nuclear issues to oil export have prevented Iran from entering Caspian projects. In 2009, Iran suggested a new energy project model that would increase the decision-making power of private companies, while the governments would become project custodians, with their decision-making power reduced (Khaghani, 2009). However, in the current political climate, the ideas proposed by Iran do not seem viable on the international level.

Nevertheless, after the events in Ukraine (addressed in detail in Chapter 5), the European Union started to reconsider its previous avoidance of Iran and plan to possibly import Iranian natural gas, with some limitations. Iran already sells about 10 billion cubic meters of natural gas per year to Turkey. If the European Union signs a contract with Iran, Iran will build a new $6-billion natural gas pipeline from South Pars to the Turkish border. The future of this deal will mainly depend on the outcome of current nuclear negotiations between Iran and the P5+1 group of countries (China, France, Russia, UK, the United States, and Germany). In negotiations with Iran, natural gas trade has been directly connected to security questions. Currently, the European Union has a ban on Iranian natural gas, issued in 2012 ('Iran's Gas Delivery to EU Possible, but with Preconditions', 2015).

Initially, the United States considered the Armenian route for the pipeline, as an attempt to solve the Armenia–Azerbaijan conflict, using the concept of pipelines as peace-building projects. The Armenian route, however, was completely rejected by Baku. It was finally decided that the pipeline would go through Georgia instead. The former Azerbaijani Ambassador to Iran, Nasib Nassibli recalls that fifteen American oil companies were invited to the White House for a special meeting, during which they were persuaded that the BTC route 'was preferable to others from the geo-strategic and geopolitical standpoints' (Nassibli, 2004, p. 168). Even Elmar Mammadyarov (Azerbaijan's Foreign Minister) admitted that decision-making about energy projects can be highly politicized at all the stages, including planning and implementation, and the

BTC pipeline was an illustrative example of it. He also emphasized the importance of the support of the pipeline by the U.S. government (Mammadyarov, 2009b). U.S. Energy Secretary Bill Richardson was one of the signatories of the Ankara Declaration of 1998 (on behalf of the U.S.), which finalized the BTC pipeline route. In the same year, the United States created a special government position for Caspian energy diplomacy, and Richard Morningstar was appointed as the first U.S. Special Counselor (Nassibli, 2004, p. 168).

Those energy developments were accompanied by military ties: in one military exercise, the United States troops joined personnel from Azerbaijan, Georgia, Turkey, Russia, Uzbekistan, and Kyrgyzstan (Nassibli, 2004, p. 168). In 2007, Foreign Minister Mammadyarov evaluated US–Azerbaijan relations as stable and productive. Azerbaijan and the United States signed the Memorandum of Understanding on Energy Security Cooperation in the Caspian Region. In addition to that, the Azerbaijani government sent troops to Afghanistan and Iraq, demonstrating its full support of the U.S. policies in the region. In 2009, Ambassador Richard L. Morningstar was appointed again as U.S. Special Envoy for Eurasian Energy. For Azerbaijan, it was a sign that Azerbaijan was on the center stage of the U.S. foreign policy and trade agenda during the Obama administration (Mammadyarov, 2009b).

The BTC pipeline was a result of what became known as 'the policy of multiple pipelines', promoted by the U.S. government in Eurasia, which stipulates that it is better to reduce dependency on any single supplier of oil and natural gas and instead to build as many various pipelines as possible, as long as they are economically feasible. Azerbaijan has fully embraced this policy: keeping its options open and periodically selecting new commercially and strategically viable projects have become the mantra of the energy policy (Mammadyarov, 2009b). As a result, all buyers of Azerbaijani hydrocarbons (e.g., Poland, Italy, Russia, Greece) regularly negotiate with Azerbaijan about future contracts. Over the period of two decades, in addition to BTC and old Soviet-era pipelines (which could not handle an increased volume of trade), Azerbaijan has considered multiple options for new routes, such as the Trans Adriatic Pipeline (through Georgia and Turkey to Italy); Nabucco (bypassing Russia); South Stream (through Russia to Bulgaria); and the Ukrainian–Polish Sarmatia (on the basis of the Odessa–Brody oil pipeline) ('Sarmatia Likely to Endorse Feasibility Report on Euro-Asian Oil Transportation Corridor on April 24', 2009).

Some of those options have been cancelled as economically unfeasible. For example, Russia recently dropped the idea of the South Stream pipeline, instead switching to the Turkish Stream project idea ('Russia/Turkey: Russia-proposed Turkish Stream Gas Project Raises Interest in Region – Turkish minister', 2015). According to the U.S. official sources, the South Stream pipeline was 'stopped[…] because the European Union[…] required that the possible future pipeline adhere to rules of competitiveness' ('House Foreign Affairs Subcommittee on Europe, Eurasia and Emerging Threats Hearing', 2015). The Russian government's official version, announced by Gazprom CEO Alexei Miller, confirmed that the European

Union 'had objected to the $40 billion South Stream pipeline, which was to enter the bloc via Bulgaria, on competition grounds' ('Dr. Alexei Borisovich Miller – Chairman, Gazprom', 2015).

As Russia–U.S. relations have experienced more and more tensions, coming to the lowest point of mutual sanctions in 2014, Azerbaijan has managed to keep diplomatic working relations with both countries. In 2009, Azerbaijan signed a deal with Russia for natural gas export. The deal caused concerns in the West about the rise of Russia's influence on Azerbaijan's decisions. Vice President Dick Cheney, visiting Azerbaijan, Ukraine, and Georgia, expressed disappointment about Azerbaijan's exports of natural gas to Russia. In his view, these exports were out of fear after the 2008 Russia–Georgia war (Lucas, 2008).

Foreign Minister Mammadyarov allayed these fears, emphasizing that Gazprom had offered 'a very attractive price' for one billion cubic meters of natural gas. He stressed that this amount was still a small fraction of Azerbaijan's overall natural gas export. He was concerned at the time that the European partners had not fully committed to buying Azeri natural gas because of the internal disagreements in the European Union (Mammadyarov, 2009b). At the time, Nabucco was the most discussed potential pipeline, and Azerbaijan indicated that it was 'not committed to Nabucco but supported the project', waiting for European partners to solve disagreements among themselves (Abdullayev, 2009). By signing the deal with Russia, Azerbaijan showed to the European partners that Europe was not the only customer for Azerbaijan's energy resources. Competition among buyers helped increase the importance of Azerbaijan as an energy provider.

Simultaneously, the Russian government was supporting Azerbaijan, referring to Azerbaijan as a focal point for peace and stability in the Caucasus region. In his speech, Russia's former special envoy Victor Kalyuzhniy called Azerbaijan a young, quickly developing sovereign nation, praising Ilham Aliyev's focus on oil and gas resources, rather than on 'costly and unreliable' renewable energy (Kalyuzhny, 2009). In the West, Russia's praising Azerbaijan was perceived as Russia pursuing its own interests, such as the South Stream pipeline, which in that planning stage competed with the U.S.-backed Nabucco (Larrabee, 2009). The European countries and the United States were concerned that Azerbaijan could potentially divert its oil and gas resources from the West to pro-Russia and pro-Iran projects.

Azerbaijan has skillfully maneuvered among the interests of the West, Russia, Iran, and even Israel and China, keeping friendly relations and increasing the country's bargaining power. The American support has been crucial in this process. The U.S. government has provided informational support and guidance to American companies in the Caspian through the U.S. Commercial Service and the Overseas Private Investment Corporation (OPIC), among other means ('Export.Gov: Helping U.S. Companies to Export', 2009):

> OPIC provided USD 100 million in political risk insurance to U.S.-based financial institutions and U.S. equity partners in the Baku–Tbilisi–Ceyhan

> oil pipeline. In 2002, OPIC invested USD 50 million in Soros Investment Capital for projects targeted to all three Caucasus countries.
>
> ('Doing Business In Azerbaijan: A Country Commercial Guide for U.S. Companies', 2008, p. 47)

As former Ambassador of Azerbaijan to the United States, Hafiz Pashayev put it, '[t]he stars and stripes do indeed follow the U.S. dollar' (Pashayev, 2009, p. 113). The president of SOCAR, Rovnag Abdullayev and other business leaders and politicians have repeatedly underscored the importance of the American support for regional project implementation (Abdullayev, 2009). The government of Azerbaijan considers the regional pipelines as a guarantor of stable U.S.–Azerbaijan relations, Azerbaijan's energy independence from Russia, and the country's commitment to cooperate with the West (Pashayev, 2009, p. 115).

One issue that remains unsolved even with strong U.S.–Azerbaijan energy ties is the Azerbaijan–Armenia conflict in Nagorno-Karabakh. The U.S. government has 'shied away' from interfering with the issue, and Azerbaijan had not received as much support from the United States as the Azeri government had hoped (Pashayev, 2009, p. 111). In 2009, a U.S. representative to the United Nations voted against Azerbaijan's attempt to bring Nagorno-Karabakh back. Ambassador Pashaev also mentioned in 2009 that the U.S. government was more critical about democracy flaws and human rights abuse in Azerbaijan than about similar problems in Armenia (Pashayev, 2009, p. 123).

Nevertheless, the Nagorno-Karabakh conflict was one of the main reasons why Azerbaijan and the Western partners chose Georgia over Armenia for the BTC route. Armenia also has close political and economic ties with Russia, and Western analysts expressed fears that Russia might use Nagorno-Karabakh for the 'divide and rule' policy in the region. Others compare Nagorno-Karabakh with South Ossetia, in the wake of the Russia–Georgia conflict of 2008. However, since Russia and Nagorno-Karabakh do not share a border, such concerns have never materialized (Kjaernet, 2010, p. 154).

Georgia, which has had friendly relations with Azerbaijan and the United States, was considered a low-risk country at the time of the BTC pipeline route selection, despite Georgia's domestic political instability in the 1990s and tense relations with Russia (Shaffer, 2009a). Although Georgia's domestic situation was complicated by South Ossetia's and Abkhazia's separatism movements, the need to bypass Russia and Armenia prevailed over the potential risks of the Georgian route. As a result, the BTC pipeline strengthened Georgia's and Turkey's position as transit states, increasing their geopolitical significance, transit revenues, and stable access to energy resources (Nassibli, 2004, p. 168). However, the BTC pipeline's safety and Georgia's significance were tested in a hard way during the Russia–Georgia war of 2008.

Russia–Georgia war of 2008

Similar to the Nagorno-Karabakh issue, discussed above, Georgia had two enclaves, Abkhazia and South Ossetia, which were once created by Joseph Stalin in the Soviet period, creating ethnic conflict and secession movements in the present. In 1931, Stalin gave Abkhazia a status of an autonomous republic (ASSR) within Georgia, even though Abkhazia had been an independent nation before it had been included in the Soviet Union (Hewitt, 2009). A similar story happened to South Ossetia. The Ossetian nation was divided into two parts – South Ossetia and North Ossetia – as part of Stalin's 'divide and rule' policy. As a result, North Ossetia became a part of the Russian Soviet Federative Republic, and South Ossetia became a part of the Georgian Soviet Republic.

The 1991 referendum in Abkhazia on the renewal of the Soviet Union overwhelmingly produced the vote of support for preserving the Soviet Union. After the dissolution of the Soviet Union happened anyway, and Georgia became an independent country, South Ossetia and Abkhazia did not want to stay within Georgia. However, the international community recognized the Georgian borders that had been established by Stalin, creating the grounds for the secession movements and potential conflict (Hewitt, 2009).

For the last two decades, the situation has been complicated by the fact that many South Ossetians and Abkhazians were in fact citizens of Russia. According to the post-Soviet Russian law, any citizen who held a Soviet passport could apply and receive Russian citizenship, and thousands of people in South Ossetia and Abkhazia used this right, obtaining their Russian citizenship. This process was happening since the breakup of the Soviet Union, and it intensified closer to the Georgia–South Ossetia conflict in 2008. The European Union accused Russia of the 'mass conferral of Russian citizenship', which was one of the factors that had led to the initiation of the military actions by Georgia in August 2008, in order to bring the breakaway republics back. Russia–Georgia tensions were also building up for several years before the culminating point in 2008. Because of tensions, Russian troops were amassed on the Russian side of the Georgia–Russia borders (Lobjakas, 2009).

In addition to the troops on the Russian side, Russian peacekeepers were inside Abkhazia and South Ossetia, according to the United Nations mandate. Even before the war, in 2006, the Georgian government was already concerned that the two enclaves might break away. The Georgian parliament passed a resolution in July 2006, ordering the Russian peacekeepers to leave. This resolution stirred a heated response in Abkhazia and South Ossetia. Stanislav Lakoba, secretary of the Abkhaz Security Council, stated: 'Georgia, because it demands the withdrawal of the CIS peacekeepers from Abkhazia, must not remain a member of the Commonwealth' ('Abkhazia Slams Georgia's Attempt to Expel Peacekeepers', 2006).

Russia also held non-peacekeeping troops in the two enclaves, and the European Union called the presence of Russian troops a violation of international law (Lobjakas, 2009). Again, as early as in 2006, Russia already made an official announcement that 'if the Georgian leadership launches a military attack against our peacekeepers, Russian citizens, and if ethnic cleansing and genocide start there, Russia will not remain indifferent'. Simultaneously, the Georgian government 'vowed to bring South Ossetia and Abkhazia back under Tbilisi's control' and 'demanded the pullout of Russian troops from the regions'. The relations between Russia and Georgia were further complicated by Russia's economic sanctions on Georgian goods and services: even the postal services and transportation between the two countries were cut off. Around the same time, Georgia detained several Russian military officers who were accused of spying on Georgia ('Russia Will Act if Georgia Starts War in Abkhazia, South Ossetia: Ivanov', 2006). It was clear that the tensions could explode any moment, even though it took about two years for the military conflict to start.

Similar to the situation with the Crimea annexation, which happened when the 2014 Winter Olympic Games were still going on, the Russia–Georgia conflict started on August 8, 2008, during the 2008 Summer Olympic Games. As the events unfolded and for another year after the war, the international media and the analysts debated about who initiated the conflict: Georgia or Russia.

The Western media reported that it was Russia, and finally, a special commission appointed by the European Union in 2009 made the conclusion. The commission, entitled the Independent International Fact-Finding Mission on the Conflict in Georgia (IIFFMCG) and led by Ambassador Heidi Tagliavini, aimed at investigating 'the origins and the course of the conflict, including with regard to international law, humanitarian law and human rights' ('Presentation of Report of Independent International Fact-Finding Mission on Conflict in Georgia', 2009). Before the report, the Western media had a very limited access to the facts on the ground, and instead the media had to rely on statements that Georgian President Mikheil Saakashvili delivered daily for English-speaking audiences. The Russian government provided very few statements and explanations of the conflict, complicating the task for the media covering the conflict (English & Svyatets, 2010).

The 1,200-page IIFFMCG report concluded that the first move in the war was made by Georgia, even though both Georgia and Russia had contributed to the build-up of the conflict. The conflict started with Georgia's shelling and artillery fire on the city of Tskhinvali on the night of August 8, 2008. The Georgian government responded to the report's conclusion by saying that Georgia had taken steps to 'protect the lives of its citizens' by starting the military actions (Lobjakas, 2009).

Russia's military response to Georgia's move into South Ossetia was overwhelming. Within five days, the Georgian army was defeated, although the Russian military stopped short of entering the capital of the country, Tbilisi. On both sides, there were hundreds of casualties, although the exact number of casualties is not known because of conflicting reports (Nichol, 2012). Already in one week after the initial military actions, on August 15, 2008, Russia agreed to a peace plan, negotiated and brokered by French President Nicolas Sarkozy. On August 25, 2008, Russia officially recognized the independence of South Ossetia and Abkhazia (Nichol, 2012). Very few other nations joined Russia in recognizing the independence of the two break-away republics: among them were Venezuela and Nicaragua. Even the closest ally of Russia, President of Belarus Alyaksandr Lukashenka, refused to recognize the independence ('Belarusian President Says Not to Recognize Georgian Breakaway Regions', 2015).

As the most recent act of disapproval, in March 2015, the European Union and the United States issued a 'condemnation' of the Russia–South Ossetia alliance and integration treaty. Both Washington and Brussels issued statements saying that the treaty 'clearly violates Georgia's sovereignty and territorial integrity'. A year before that, in 2014, the United States did not recognize a similar treaty between Russia and Abkhazia ('EU, U.S. Condemn Russia, South Ossetia "Treaty"', 2015).

The Russia–Georgia war, as short as it was, created a major concern for the neighboring countries, and especially for Azerbaijan, in relation to the physical security of the BTC pipeline. The fear in Azerbaijan was that the pipeline might be damaged during the military actions, accidentally or on purpose. Fortunately, these fears did not materialize. A vice-president of BP-Azerbaijan, Khalilov, confirmed that the Russian forces had not attacked the pipeline or any other BP facilities. Although the pipeline was non-operational for several days during the war, it was, according to Khalilov, because of a terrorist attack (explosion) by the Kurdish separatists on the Turkish side. After the damage from this explosion was repaired, the pipeline continued operating normally even during the war (Khalilov, 2009).

After the Russia–Georgia war, Azerbaijan continued to balance between Russia and the United States. During the Russia-U.S. 'reset', the government of Azerbaijan was concerned that the 'reset' and the closeness between the United States and Russia meant that the United States would be turning away from the Caspian region, undermining U.S.–Azerbaijan relations. However, U.S. Secretary of State, Hillary Clinton visited Azerbaijan and Georgia in 2011 and assured the government in the U.S. commitment in 'promoting democracy and strengthening U.S. ties in the region'. In addition to the continued support of Azerbaijan's energy policy, Clinton urged President Ilham Aliyev to solve the Nagorno-Karabakh dispute. At the time, the U.S. government worked closely on encouraging a border-opening deal between Armenia and Turkey. Experts noted that the government of Azerbaijan felt that the United States did not do enough to link the deal with the Nagorno-Karabakh resolution, and that the U.S. foreign policy was 'inconsistent' because of the new closeness to Russia during the 'reset' (Champion & Fairclough, 2010).

During the 'reset', Azerbaijan continued to maintain positive relations with Moscow. One of the reasons was the Turkey–Armenia rapprochement of 2009, which did not work out in the end but looked promising in 2009. In October, Armenia and Turkey signed a deal to normalize their relations and reopen their borders (closed since 1993) ('Opening a Border: With Help from Hillary Rodham Clinton, Turkey and Armenia Take a Step Toward Rapprochement', 2009). Because of this development, Azerbaijan started to consider the energy routes that would include Russia, rather than bypassing it, in order to reduce the reliance on Turkey and to diversify transit options.

As a result, economic and trade relations between Azerbaijan and Russia improved by 2011, which was emphasized by Azerbaijan's Foreign Minister Elmar Mammadyarov during his visit to Moscow. He came to Moscow to negotiate with his Russian counterpart, Sergey Lavrov, about the increasing export of Azeri oil (via the Baku–Novorossiysk pipeline) and natural gas to Russia. SOCAR and Gazprom signed another natural gas export contract, which again stirred concerns in the West. This contract was followed by an additional agreement that allowed Gazprom to import 2 billion cubic meters of Azeri natural gas per year ('Azerbaijan, Russia Discuss Increasing Oil and Gas Export', 2011). Some experts described those deals between Russia and Azerbaijan as the 'economization', i.e. a policy that prioritized the economic and business factors in decision-making (Kjaernet, 2010, p. 151).

The deals with Russia caused the U.S.–Azerbaijan relations to slightly cool off. As a result, the activities and meetings of the U.S.–Azerbaijan Intergovernmental Commission on Economic Cooperation slowed down and were resumed only in 2012 (Nichol, 2014). Nevertheless, Azerbaijan remained a 'strategic partner' for the United States in U.S. counter-terrorism operations, participating in NATO's Northern Distribution Network, in order to bring U.S. supplies to and from Afghanistan. Azerbaijan has also actively promoted the Southern Corridor for natural gas transportation from Central Asia and the Caucasus regions to Western Europe. The Southern Corridor has also been supported by the United States in energy security of the region (Nichol, 2014).

In sum, Azerbaijan has balanced between the United States and Russia, in order to ensure the country's independence in energy export decision-making. The West has been interested in bypassing Russia and reducing its influence in the region, because of the geopolitical rivalries and calculations. The BTC pipeline route was probably the most expensive and technologically challenging among the other options but it bypassed Russia and Iran and went through friendly Georgia and Turkey. Other pipelines will follow suit.

Domestic interest groups

One can predict that if powerful domestic lobbies are in favor of energy cooperation, the chances of cooperation are higher. If powerful lobbies are divided on some issue or focused on other issues, they will have less power. In the case of Azerbaijan–U.S. cooperation on energy issues, both governments have very tightly controlled the process, so the room for maneuvering by domestic lobbies has not been as large. One powerful lobby that has been especially notable in this case is the Armenian lobby in the United States. Because of the Armenia–Azerbaijan conflict, the Armenian lobby tried to slow down the U.S.–Azerbaijan energy ties. Such associations as the Armenian Cultural Association of America, the Armenian National Committee of America, the Armenian Assembly of America, the Armenian General Benevolent Union, the Armenian Church, and others in the United States have been working on multiple campaigns in favor of Armenia. Azerbaijan, however, as well as Georgia, has been more successful in the ties with the U.S. government to promote favorable policies toward the Caspian region (MacFarlane, 2011, p. 112).

The Armenian community has criticized the political system in Azerbaijan and the way Azerbaijan's energy sector is tightly controlled by the state. It emphasized that the power of the Aliyev family, which has ruled in Azerbaijan 'for all but six years since 1969', has been growing, the country has not developed fully functioning democratic institutions – all in accordance with the predictions of the resource curse literature that argues that oil-dependent states suffer from the lack of democracy, human rights issues, and the underdevelopment of non-oil industries. Petroleum wealth allows such governments to 'stay in power simply by doling out patronage – whether through subsidies, tax cuts, public sector employment, or lucrative contracts' and to 'buy political support, pay off opposition, and depoliticize the population (Overland, Kendall-Taylor, & Kjaernet, 2010, p. 4).

From the early stages of the Armenia–Azerbaijan conflict over Nagorno-Karabakh in the 1990s, the Armenian community has lobbied for U.S. support of Armenia. Initially, some of those efforts were successful. For example, in the 1990s, the United States cut off the U.S. economic assistance to Azerbaijan, under Section 907 of the Freedom Support Act, when Armenia was using U.S. support for demanding that Azerbaijan lift the economic blockade of Nagorno-Karabakh ('Nagorno-Karabakh', 2009).

In 2001, when the BTC pipeline route was in discussion, the Armenian lobby asked U.S. President George W. Bush to facilitate the route for a pipeline from Azerbaijan to Turkey via Armenia, instead of the BTC route. Armenia even proposed a plan that would make the pipeline $600 million cheaper than the current BTC pipeline ('Armenian Lobby Applies Pressure to Get BTC Passing Through Armenia', 2001). However, this attempt was not successful, and the BTC route was selected.

A decade later, in 2011, Armenian lobbyists influenced U.S. Congress members to withdraw the U.S. Ambassador to Azerbaijan, Matthew Bryza, who had held a temporary ambassadorial position in Baku since 2010. Previously, Bryza had been the Deputy Assistant Secretary of State for European and Eurasian Affairs and an advisor to the U.S. government on Eurasian energy issues. He was one of the negotiators in the Nagorno-Karabakh mediation talks and a multiple pipelines policy architect who helped shape the role of Azerbaijan in the Caspian energy security and energy supplies to the West (Lomsadze, 2012).

The Armenian community claimed that the Ambassador had excessively close ties to Azerbaijan and Turkey. His wife, Zeyno Baran, is a Turkish American and one of the leading scholars in the United States on Turkish foreign policy. She has worked as the Director of the Center for Eurasian Policy and a Senior Fellow at the Hudson Institute – both powerful think tanks in Washington, DC. Influenced by the large and influential Armenian community in the United States, senators Barbara Boxer of California and Robert Menendez of New Jersey were vocal opponents of Ambassador Bryza's confirmation. Both senators had well-known public ties with the Armenian community in the United States (Aliyev, 2012).

Reaction by the government of Azerbaijan to senators Boxer and Menendez's attempt to block the Ambassador's confirmation was very vocal. Azerbaijani leaders such as Eldar Namazov, Chief of Staff under the late President Heydar Aliyev; Leyla Aliyeva, Center for National and International Studies Director; Ilgar Mammadov, Political Analyst; Sabit Bagirov, Economic Analyst; and Mehman Aliyev, Turan News Agency Director; Elhan Shahinoglu, Head of Baku's Atlas research center all protested in response. Azerbaijan's Deputy Foreign Minister Araz Azimov said that this 'could become a bad precedent for American diplomats who would not know whose policy they should pursue – the president's policy, the Senate's or the interests of a handful of lobbyists' (Abbasov, 2012). He implied that the Armenian lobby had strong influence on the senators. In 2009, Sen. Robert Menendez had received an award from the Armenian National Committee of America (ANCA) for his commitment to the Armenian American community. When giving the award to the senator, the ACAA Board Member George Aghjayan praised his support of Armenia on the issues of the Armenian Genocide and Nagorno-Karabakh ('Sen. Robert Menendez to Receive 2009 ANCA-ER Freedom Award at Third Annual Banquet', 2009).

The withdrawal of the U.S. Ambassador to Azerbaijan was not the only political process that the Armenian-American community was participating in. Another one was a strong opposition to an Armenia–Turkey reconciliation agreement. The agreement between the two countries about the re-opening of the border between them was signed in 2009. The border had been closed by Turkey in 1993 in solidarity with Azerbaijan after Armenian troops had occupied Nagorno-Karabakh. However, under pressure from several circles, the most prominent of which was the Armenian lobby in the United States, the Armenian government pulled the plug on the 2009 agreement in the same year ('Turkish Institute for Progress Applauds Armenian President Serzh Sargsyan for Joining Call for a Progressive Approach to Turkish-Armenian Relations', 2015).

The Armenia–Turkey rapprochement process of 2008–2009 was blocked by the Armenian lobby partly because of the Nagorno-Karabakh issue. Armenian experts, such as military and political analyst Richard Giragosian and others, were disappointed about how the rapprochement process developed: by 2011, they noticed that the process was no longer 'a priority for Turkey'. More importantly, Turkey's omission of the resolution on Nagorno-Karabakh in the rapprochement process was unacceptable for Armenia:

> ...there is no reference to Karabakh within the Protocols. And to attempt at this late stage to re-link the issues is unhelpful at best and insincere at worst. So what we see in Turkey seems rather insincere and in danger of being perceived as an unreliable and unready interlocutor for Armenia.
>
> ('Armenia Delegation Disappointed with Istanbul Symposium to Revitalize Armenia-Turkey Rapprochement: Giragosian', 2011)

The powerful Armenian Diaspora in the United States held a position on the Turkey–Armenia border reopening, which was very different from what Armenian President Serzh Sargsyan was advocating for. When Sarkisian came to Glendale, California (the city with the largest Armenian population outside of Armenia) in 2009 to discuss the reconciliation protocols, about 10,000 Armenians protested Sarkisian's agenda (Khachatourian, 2009). The protestors felt that the President came to 'force [the protocols] down our throats', as a local Armenian commentator wrote in an Armenian American publication. The Turkey–Armenia agreement was harshly criticized by the Armenian Diaspora as 'dangerous and defeatist documents that are sure to irreversibly change the course of Armenian history' (Khachatourian, 2009).

Critics of the protocols also argued that they did not recognize the Armenian Genocide ('Turkey, Armenia Agree to Forge Ties', 2009). This has been the most debated issue between the two countries, especially in 2015, when it was a 100-year anniversary of the genocide (as it is claimed by the Armenian community), or mass killings (as it is claimed by Turkey, without using the word 'genocide'). Senator Menendez, who played a key role in Ambassador Bryza's story, mentioned above, called the protocol provisions, offered by Turkey, 'frankly absurd', and 'an insult to the Armenian people'. Referring to the absence of the Armenian Genocide issue in the protocols, Menendez continued about the protocols: 'It is an insult to the memory of the victims. It is an insult to the top scholars who have clearly spoken on this issue and issued their unequivocal conclusions' ('Sen. Menendez Calls Protocols "Insult" to Armenian Nation', 2009).

Only recently, there has been some progress in Turkey–Armenia relations, as Armenian President Serzh Sargsyan announced on April 22, 2015: 'We shall at the end of the day establish normal relations with Turkey and establishing those normal relations should be without any preconditions'. American politicians have also expressed support for this progress in the Turkey–Armenia reconciliation. Former Representative Solomon Ortiz (D-TX) said: 'The actions taken by Presidents Sargsyan, Erdogan, and Obama mark a new chapter in Turkish–American relations, one defined by peace and progress, which is something nearly every American supports' ('Turkish Institute for Progress Applauds Armenian President Serzh Sargsyan for Joining Call for a Progressive Approach to Turkish–Armenian Relations', 2015).

However, the issue of Nagorno-Karabakh is still impeding the possibility of Turkey–Armenia reconciliation. Nagorno-Karabakh held the Armenia-controlled parliamentary elections on May 3, 2015. For Armenia, the elections were an expression of democracy in the enclave. For the critics, such as Azerbaijan and Turkey, they were another impediment in a possible resolution of this frozen conflict. Turkish Foreign Ministry stated about the elections:

> We condemn this step, which is a violation of the sovereignty and territorial integrity of Azerbaijan, and don't recognize these illegitimate elections. As a member of the OSCE Minsk Group, Turkey will continue to support

> efforts aimed at finding a fair and lasting solution to the Nagorno-Karabakh conflict.
>
> ('Turkey: Nagorno-Karabakh Election Will Harm Peace Efforts', 2015)

These examples of the Armenian Diaspora's activities show how powerful and influential it is in the United States. However, it was not able to block the U.S.–Azerbaijan energy contracts: the lobby was not able prevent the construction of the BTC pipeline or to divert its route via Armenia. The reason was that energy policy decisions have been closely controlled by the governments of the United States, Azerbaijan, Turkey, and other participating states. The geopolitical considerations prevailed in the BTC pipeline selection. Simultaneously, non-energy issues, such as the Nagorno-Karabakh conflict and the Armenian Genocide, have been the center of attention for the Armenian Diaspora. The lobby has been more united and influential in those aspects than in the oil and gas issues.

Azerbaijan: lessons for the future

This chapter explained the energy ties between the United States and Azerbaijan, as well as the strong American support of Azerbaijan's position in determining the BTC pipeline route and other multibillion dollar projects. The pipeline was an interesting case to explore for several reasons: the geographic distance of Azerbaijan from the United States, varying estimations of Azerbaijan's oil and gas reserves, and Azerbaijan's domestic institutions and friendly relations with Iran, among others.

Economic factors can partially explain why Azerbaijan has been so successful in cooperating with the West. The investment climate in Azerbaijan has been substantially improved over the last two decades, and the government of Azerbaijan has provided very strong guarantees to Western investors, compared to other post-Soviet states. However, the investment climate alone does not explain some of the aspects of the energy deals, such as the selected route of the BTC pipeline.

When the Baku–Tbilisi–Ceyhan pipeline route was being determined, the geopolitical considerations by all the involved governments played the ultimate role, even above the purely economic calculations of the pipeline's financial costs and technical options. The proposed alternative routes – via Russia, Armenia, Iran – might have been much cheaper and technologically easier, but they did not deliver the main task promoted by the policy of multiple pipelines: to provide Azerbaijani oil and gas to NATO partners in Europe via friendly transit countries (in this case, Georgia and Turkey). The state control of the Azerbaijani energy industry and a close guidance by the U.S. government of American energy companies were also in line with those geopolitical considerations.

The domestic interest groups that were opposing BTC in this case were not strong enough to modify the strategies chosen by the governments. The powerful Armenian lobby in the United States was able to block the re-appointment of the U.S. Ambassador to Azerbaijan and the Turkey–Armenia rapprochement. However, it did not have the power to block BTC or to divert the pipeline's route to Armenia.

Despite a remarkable success with hydrocarbon exports, the most important goals for Azerbaijan at the moment are to continue diversifying the country's economy and building democratic institutions. Azerbaijan has already started investing in non-oil sectors. In 2013, the volume of non-oil industries expanded by almost 10 per cent, compared to the level one year before that ('Azerbaijan: Vested Interests Will Resist Change', 2014). Oil and natural gas are finite resources, and investing the revenues into the non-oil sectors, such as services, transport, and trade, will pay off in the future. Finally, climate change that we are experiencing globally is already a push for all governments to reform their energy sector, and Azerbaijan needs to be one of them.

References

18 October – The National Independence Day. (2009). Retrieved September 22, 2009, from http://www.president.az/browse.php?sec_id=55#18

Abbasov, S. (2012). Azerbaijan: Baku Fumes Over Scuttled Ambassadorial Appointment. *EurasiaNet.Org*, (January 5). Retrieved from http://www.eurasianet.org/node/64796

Abdullayev, R. (2009, July 4). Meeting with the Energy Politics and Economics Summer School Participants.

Abkhazia Slams Georgia's Attempt to Expel Peacekeepers. (2006, July 18). *Daily News Bulletin*, p. 1.

Alieva, L. (2011). Regional Cooperation in the South Caucasus: The Azerbaijan Perspective. In M. Aydin (Ed.), *Non-Traditional Security Threats and Regional Cooperation in the Southern Caucasus* (pp. 197–207). Amsterdam: IOS Press.

Aliyev, M. (2012, June 21). Top Official: Appointment of U.S. Ambassador to Azerbaijan Necessary for Bilateral Relations. *McClatchy – Tribune Business News.*

Armenia Delegation Disappointed with Istanbul Symposium to Revitalize Armenia–Turkey Rapprochement: Giragosian. (2011, January 11). *Press.Am.* Retrieved from http://www.epress.am/en/2011/11/01/armenia-delegation-disappointed-with-istanbul-symposium-aimed-at-revitalizing-armenia-turkey-rapprochement-giragosian.html

Armenian Lobby Applies Pressure to Get BTC Passing Through Armenia. (2001, July 28). *Hurriyet Daily News.* Retrieved from http://www.hurriyetdailynews.com/default.aspx?pageid=438&n=armenian-lobby-applies-pressure-to-get-btc-passing-through-armenia-2001-07-28

Azerbaijan. (2006). *International Financial Statistics Yearbook*, *47*, 912.

Azerbaijan, Armenia Seek End to War. (1992, February 26). *South China Morning Post & the Hong Kong Telegraph*, p. 11.

Azerbaijan: Country Fact Sheet. (2015, May 1). *EIU ViewsWire.*

Azerbaijan: Country Outlook. (2010). New York: The Economist Intelligence Unit N.A., Incorporated.

Azerbaijan Petrochemicals Report – Q2 2015. (2015) (pp. 1–60). London: Business Monitor International.

Azerbaijan, Russia Discuss Increasing Oil and Gas Export. (2011). *Oil and Gas Eurasia.* Retrieved from http://www.oilandgaseurasia.com/news/p/0/news/12124

Azerbaijan: Vested Interests Will Resist Change. (2014) (pp. 1–n/a). Oxford: Oxford Analytica Ltd.

Azerbaijan.az. History of SOCAR. Retrieved September 23, 2009, from http://www.azerbaijan.az/portal/StatePower/Committee/committeeConcern_01_e.html

Azeri MP Urges More Attention to Problems of Ethnic Azeris in Iran. (2014, September 14). *BBC Monitoring Central Asia*.

Bagirov, H. (2009, July 10). Lecture to the Energy Politics and Economics Summer School Participants.

Belarusian President Says Not to Recognize Georgian Breakaway Regions. (2015, April 23). *BBC Monitoring Former Soviet Union*.

Blanchard, R. D. (2005). *The Future of Global Oil Production: Facts, Figures, Trends and Projections, by Region*. Jefferson, NC: McFarland & Company, Inc.

BP Becomes Stakeholder in Azeri-led Regional Gas Project. (2015, April 25). *BBC Monitoring Central Asia*.

Champion, M., & Fairclough, G. (2010, July 05). Europe News: U.S. Seeks to Warm Ex-Soviet Ties. *Wall Street Journal*, p. 5.

Dabrowski, M. (2013). Monetary Policy Regimes in CIS Economies and their Ability to Provide Price and Financial Stability. *BOFIT Discussion Papers, 2013*(8), 4–53.

Doing Business 2009. (2008) (pp. 1–211). Washington, DC: The International Bank for Reconstruction and Development / The World Bank.

Doing Business in Azerbaijan: A Country Commercial Guide for U.S. Companies. (2008). Washington, DC: U.S. Commercial Service.

Doing Business in the New Independent States – A Resource Guide. (1994) (July 1994 ed., Vol. 5, pp. 497–511): U.S. Department of State Dispatch.

Dr. Alexei Borisovich Miller – Chairman, Gazprom. (2015). San Francisco: Boardroom Insiders, Inc.

Eastern Europe Politics: Construction Begins of Trans-Anatolian Pipeline. (2015). New York: The Economist Intelligence Unit N.A., Incorporated.

Emerging Europe Oil and Gas Insight – May 2015. (2015) (pp. 1–9). London: Business Monitor International.

English, R., & Svyatets, E. (2010). *A Presumption of Guilt? Western Media Coverage of the 2008 Russia–Georgia War*. Los Angeles, CA: University of Southern California.

EU, U.S. Condemn Russia, South Ossetia 'Treaty'. (2015). Lanham: Federal Information & News Dispatch, Inc.

Export.Gov: Helping U.S. Companies to Export. (2009): U.S. Government.

Finn, R. (2009). Diplomatic Beginning in Baku. In A. Petersen & F. Ismailzade (Eds.), *Azerbaijan in Global Politics: Crafting Foreign Policy* (pp. 97-105). Baku, Azerbaijan: Azerbaijan Diplomatic Academy.

Guliyev, F. (2013). Oil and Regime Stability in Azerbaijan. *Demokratizatsiya, 21*(1), 113–147.

Guseinov, V. (2008). *Politica u Poroga Tvoego Doma. Izbrannie Publitsisticheskie i Analiticheskie Statji*. Moscow: Krasnaya Zvezda.

Hewitt, G. (2009). Abkhazia, Georgia, and the Crisis of August 2008: Roots and Lessons. *Global Dialogue, 11*, 11–22.

Heydar Alirza oglu Aliyev (2009). Retrieved September 23, 2009, from http://aliyev-heritage.org/en/biography2.html

House Foreign Affairs Subcommittee on Europe, Eurasia and Emerging Threats Hearing. (2015). Lanham, MD: Federal Information & News Dispatch, Inc.

Howard Berman Writes Secretary of State Clinton, Calls for Azerbaijan's Suspension from NATO Partnership for Peace Program and Ending Arms Sales to Azerbaijan. (2012). Lanham: Federal Information & News Dispatch, Inc.

Ibadoglu, G. (2011). Economics of Natural Resources and Transition in Southern Caucasus. In M. Aydin (Ed.), *Non-Traditional Security Threats and Regional Cooperation in the Southern Caucasus* (pp. 3–16). Amsterdam: IOS Press.

Iran's Gas Delivery to EU Possible, but with Preconditions. (2015, May 2). *McClatchy – Tribune Business News.*

Ismayilov, E. (2014, October 11). State Oil Fund of Azerbaijan Expects Balanced Budget in 2014. *McClatchy – Tribune Business News.*

Kalyuzhny, V. (2009). *Keynote Remarks.* Paper presented at the Azerbaijan in Global Politics: Crafting Foreign Policy, Baku, Azerbaijan.

Kendall-Taylor, A. (2010). Introduction: Domestic Balancing Acts in the Caspian Petro-States. In I. Overland, H. Kjaernet, & A. Kendall-Taylor (Eds.), *Caspian Energy Politics* (pp. 15–19). London: Routledge.

Khachatourian, A. (2009, October 2). A Gala for the Protocols. *Asbarez.Com.* Retrieved from http://asbarez.com/71335/a-gala-for-the-protocols/

Khaghani, M. (2009, July 6). Lecture to the Energy Politics and Economics Summer School Participants.

Khalilov, S. (2009, July 5). Lecture to the Energy Politics and Economics Summer School Participants.

Khalilova, I. (2013, October 29). Azerbaijan Improves Position in 'Doing Business – 2014' Rating. *McClatchy – Tribune Business News.*

Kjaernet, H. (2010). Azerbaijani–Russian Relations and the Economization of Foreign Policy. In I. Overland, H. Kjaernet, & A. Kendall-Taylor (Eds.), *Caspian Energy Politics: Azerbaijan, Kazakhstan and Turkmenistan* (pp. 150–161). London: Routledge.

Larrabee, S. (2009, July 15). *Emerging Russian–Turkish Relations.* Paper presented at the Turkey, Russia, and Regional Energy Strategies, Washington, DC.

Lee, T. (2006). The Extractive Industries Transparency Initiative: Public Information on Natural Resource Revenues. *Petroleum Accounting and Financial Management Journal, 25*(2), 1–18.

LeVine, S. (2001, August 3). Sea or Lake? Hunt for Caspian Oil Stokes Border Feuds And Arcane Theories – Iranian Threat Adds Pressure To Resolve Many Claims And Maps of Five Nations – A Need to Know 10 Treaties. *Wall Street Journal.*

Lobjakas, A. (2009, October 11). News Analysis: EU Report on 2008 War Tilts against Georgia. *Ukrainian Weekly*, pp. 3, 19.

Lomsadze, G. (2012). Matthew Bryza: From Oil Policy to the Oil Business. *Eurasianet.Org.* Retrieved from http://www.eurasianet.org/node/65524

Lucas, E. (2008, September 8). How the West is Losing the Energy Cold War. *The Times (London)*, p. 28.

MacFarlane, N. (2011). The Evolution of US Policy towards the Southern Caucasus. In M. Aydin (Ed.), *Non-Traditional Security Threats and Regional Cooperation in the Southern Caucasus* (pp. 107–124). Amsterdam: IOS Press.

Maharramov, R. (2010). Petroleum-Fuelled Public Investment in Azerbaijan: Implications for Competitiveness and Employment. In I. Overland, H. Kjaernet, & A. Kendall-Taylor (Eds.), *Caspian Energy Politics: Azerbaijan, Kazakhstan and Turkmenistan* (pp. 38–59). London: Routledge.

Mammadov, I. (2010). *The Geopolitics of Energy in the Capian Sea Region.* Lewinston, NY: The Edwin Mellen Press.

Mammadov, N. (2009). Azerbaijan's Relations with the Islamic World and the Countries of Asia. In A. Petersen & F. Ismailzade (Eds.), *Azerbaijan in Global Politics: Crafting Foreign Policy* (pp. 149–171). Baku, Azerbaijan: Azerbaijan Diplomatic Academy.

Mammadyarov, E. (2009a, July 9). *Keynote Remarks*. Paper presented at the Azerbaijan in Global Politics: Crafting Foreign Policy, Baku, Azerbaijan.

Mammadyarov, E. (2009b, July 5). Meeting with the Energy Politics and Economics Summer School Participants.

Mann, S. R. (2003). Caspian Futures. In J. H. Kalicki & E. K. Lawson (Eds.), *Russian-Eurasion Renaissance? U.S. Trade and Investment in Russia and Eurasia* (pp. 147–171). Washington, DC: Woodrow Wilson Center Press.

Mattson, G. (2009, July 10). Lecture to the Energy Politics and Economics Summer School Participants.

Mehdiyeva, N. (2011). *Power Games in the Caucasus: Azerbaijan's Foreign and Energy Policy towards the West, Russia and the Middle East*. London: I.B.Taurus.

Mortished, C. (2008, September 3). Europe Has Weak Hand in a Poker Game of Power. *The Times (London)*, p. 49.

Nagorno-Karabakh. (2009, September 23). *GlobalSecurity.Org*. Retrieved from http://www.globalsecurity.org/military/world/war/nagorno-karabakh.htm

Nanay, J. (2009). Russia's Role in the Eurasian Energy Market: Seeking Control in the Face of Growing Challenges. In J. Perovic, R. Orttung, & A. Wenger (Eds.), *Russian Energy Power and Foreign Relations: Implications for Conflict and Cooperation* (pp. 109–131). London: Routledge.

Nassibli, N. (2004). Azerbaijan: Policy Priorities Towards the Caspian Sea. In S. Akiner (Ed.), *The Caspian: Politics, Energy, and Security* (pp. 157–180). New York, NY: Routledge Curzon.

New Azerbaijan–EU Agreement to Help Expand Trade. (2015, March 30). *McClatchy – Tribune Business News*.

New Well Within Largest Gas Project Drilled in Azerbaijan. (2015, April 20). *McClatchy – Tribune Business News*.

News in Brief. (2009, February 1). *Petroleum Economist*. Retrieved from http://www.proquest.com.libproxy.usc.edu/

Nichol, J. (2012). Russia–Georgia Conflict in August 2008: Context and Implications for U.S. Interests. *Russia, China and Eurasia, 28*(1), 1–49.

Nichol, J. (2014). Armenia, Azerbaijan, and Georgia: Political Developments and Implications for U.S. Interests. *Current Politics and Economics of Russia, Eastern and Central Europe, 29*(2), 193–279.

Opening a Border: With Help from Hillary Rodham Clinton, Turkey and Armenia take a Step Toward Rapprochement. (2009, Oct 14). *The Washington Post*.

Overland, I., Kendall-Taylor, A., & Kjaernet, H. (2010). Introduction: The Resource Curse and Authoritarianism in the Caspian Petro-States. In I. Overland, H. Kjaernet, & A. Kendall-Taylor (Eds.), *Caspian Energy Politics: Azerbaijan, Kazakhstan and Turkmenistan* (pp. 1–11). London: Routledge.

Papava, V. (2011). Economic Transformation and the Impacts of the Global Financial Crisis in the Southern Caucasus. In M. Aydin (Ed.), *Non-Traditional Security Threats and Regional Cooperation in the Southern Caucasus* (pp. 17–20). Amsterdam: IOS Press.

Pashayev, H. (2009). Azerbaijan–U.S. Relations: From Unjust Sanctions to Strategic Partnership. In A. Petersen & F. Ismailzade (Eds.), *Azerbaijan in Global Politics: Crafting Foreign Policy* (pp. 109–129). Baku, Azerbaijan: Azerbaijan Diplomatic Academy.

Presentation of Report of Independent International Fact-Finding Mission on Conflict in Georgia. (2009, October 5). *US Fed News Service, Including US State News*.

Razavi, H. (2007). *Financing Energy Projects in Developing Countries*. Tulsa, OK: Penn Well Corporation.

Razumovskaya, O. (2010, July 28). Hayward to 'Get His Life Back' in Russia. *The Moscow Times*. Retrieved from http://www.themoscowtimes.com/news/article/hayward-to-get-his-life-back-in-russia/411178.html

Russia. (2010, September 20). *Russia & CIS Business & Financial Daily*. Retrieved from http://www.lexisnexis.com.libproxy.usc.edu/hottopics/lnacademic/?

Russia Scraps South Stream Natural Gas Pipeline Project. (2014, December 2). *Progressive Digital Media Oil & Gas, Mining, Power, CleanTech and Renewable Energy News*.

Russia Will Act if Georgia Starts War in Abkhazia, South Ossetia: Ivanov. (2006, October 8). *Xinhua News Agency – CEIS*, p. 1.

Russia/Turkey: Russia-proposed Turkish Stream Gas Project Raises Interest in Region – Turkish minister. (2015, April 10). *Asia News Monitor*.

Sabonis-Helf, T. (2009, July 9). European Energy Security and Russian/Caspian Energy Supply. Lecture for the Energy Politics and Economics Summer School Participants.

Sadigov, V. (2009). Foreign Policy of Azerbaijan: European Dimension. In A. Petersen & F. Ismailzade (Eds.), *Azerbaijan in Global Politics: Crafting Foreign Policy* (pp. 131–148). Baku, Azerbaijan: Azerbaijan Diplomatic Academy.

Sagheb, N., & Javadi, M. (1994). Azerbaijan's 'Contract of the Century' Finally Signed with Western Oil Consortium. *Azerbaijan International*, *2*(4), 26–28, 65. Retrieved from http://azer.com/aiweb/categories/magazine/24_folder/24_articles/24_aioc.html

Sallis, D. (2003). Azerbaijan: Gateway to the Newly Independent States. *Euroinvest*, 50–52.

Sarmatia Likely to Endorse Feasibility Report on Euro-Asian Oil Transportation Corridor on April 24. (2009, April 9). Retrieved August 17, 2009, from http://www.kyivpost.com/world/39287

Sen. Menendez Calls Protocols 'Insult' to Armenian Nation. (2009, October 3). *Asbarez.Com*. Retrieved from http://asbarez.com/71445/sen-menendez-calls-protocols-insult-to-armenian-nation/

Sen. Robert Menendez to Receive 2009 ANCA-ER Freedom Award at Third Annual Banquet. (2009, September 11). *The Armenian Weekly*. Retrieved from http://www.armenianweekly.com/2009/09/11/sen-robert-menendez-to-receive-2009-anca-er-freedom-award-at-third-annual-banquet/

Shaffer, B. (2009a, July 5). Lecture to the Energy Politics and Economics Summer School Participants.

Shaffer, B. (2009b). Permanent Factors in Azerbaijan's Foreign Policy. In A. Petersen & F. Ismailzade (Eds.), *Azerbaijan in Global Politics: Crafting Foreign Policy* (pp. 67–84). Baku, Azerbaijan: Azerbaijan Diplomatic Academy.

Shah Deniz Signed: Third Major Contract in Caspian. (1996). *Azerbaijan International*, 4(3), 35. Retrieved from http://azer.com/aiweb/categories/magazine/43_folder/43_articles/43_shah.html

Suleymanov, E. (2009, July 09). *Security and Conflict in the Region: Finding a Careful Balance of Power*. Paper presented at the Azerbaijan in Global Politics: Crafting Foreign Policy, Baku, Azerbaijan.

Tariverdiyeva, E. (2015, May 1). OSCE MG urges Azerbaijan, Armenia to Intensify Talks. *Trend News Agency*. Retrieved August 25, 2015, from http://en.trend.az/azerbaijan/karabakh/2390033.html

Turkey, Armenia Agree to Forge Ties. (2009, October 10). *CBS News World*. Retrieved from http://www.cbc.ca/news/world/story/2009/10/10/turkey-armenia.html

Turkey: Nagorno-Karabakh Election Will Harm Peace Efforts. (2015, April 30). *Anadolu Agency: AA*.

Turkish Institute for Progress Applauds Armenian President Serzh Sargsyan for Joining Call for a Progressive Approach to Turkish-Armenian Relations. (2015, April 23). *PR Newswire*.

Visit to Baku, Azerbaijan, Jan 22–24 October 1998. (1999). Retrieved September 23, 2009, from http://www.nato-pa.int/archivedpub/trip/as65pced-baku.asp

Yergin, D. (2011). *The Quest: Energy, Security, and the Remaking of the Modern World*. New York, NY: Penguin Books.

Zygar, M., & Grib, N. (2009). Mister Neft. *Kommersant, 78/P*(4133). Retrieved from http://kommersant.ru/doc.aspx?DocsID=1164267&NodesID=4

4 Russia–Germany energy cooperation

Introduction

Russia and Germany have closely cooperated in the oil and natural gas trade and investment for many decades, even during the Soviet time. Even though Germany is a leading state in the European Union, the German government has not always followed the European policy of reducing imports of oil and gas from Russia in favor of other suppliers. Instead, Germany and Russia built the Nord Stream – the $10.2-billion, 761-mile-long pipeline – on the bottom of the Baltic Sea, directly connecting the two countries ('Germany, Russia to Launch Nord Stream', 2011).

Nord Stream has been harshly criticized by the EU and the United States. The pipeline bypassed Poland and the Baltic transit states, which had their influence and their transit fees reduced. The pipeline also stirred a heated debate in the European Union about an increased dependency on Russian oil and gas: 'the German government … has been the prime example of a European country actively seeking to foster dependencies with Russia' (Perovic, Orttung, & Wenger, 2009, p. 11). The Bush Administration in the United States was also strongly against this pipeline, because the administration supported the interests of the Baltic States and Ukraine. The U.S. government was promoting the policy of multiple pipelines that was aimed at reducing the energy trade between Europe and Russia. The U.S. Ambassador to Sweden, Michael Wood, called the Nord Stream a 'special arrangement between Germany and Russia', instead advising the European Union to focus on developing a more unified energy policy (Chazan, 2009).

The Ambassador's words reflected a common theme in the European Union, in which countries such as Poland and the Baltic States had been working on reducing the dependency on Russian natural gas. Critics argued that the pipeline at the bottom of the Baltic Sea was not economically justifiable, and that the land option would be much cheaper. Finally, environmentalists were concerned with the pipeline's effect on the ecosystem of the Baltic Sea. Nevertheless, Germany and Russia proceeded with the construction, and the pipeline is now in operation.

In response to all criticism, the German and Russian governments repeatedly stated that the pipeline on the bottom of the sea was the most economical and environmentally friendly way to deliver natural gas to Europe. They showed economic calculations that the pipeline would pay for itself very quickly, delivering natural gas at a competitive price. The President of Russia at the time, Dmitry Medvedev, stated at the opening ceremony of the pipeline:

> Natural gas from Russia and the energy produced from this gas will help to ensure reliable electricity supply to European consumers and thus strengthen energy security and improve the lives of so many people, which is probably what matters the most to all of us.
>
> ('Medvedev Speaks at Nord Stream Gas Pipeline Launch Ceremony – Kremlin Transcript', 2011)

The implementation of the Nord Stream pipeline, while so many other pipeline projects in Europe have never materialized, is an interesting case to explore for several reasons. First, the investment climate for foreign businesses in Russia has been increasingly difficult over the last decade. Even the Russian government admitted in 2010 (when Nord Stream was being built) that Russia's 'Doing Business' rating was 120th out of 183 countries in terms of business conditions. Among the factors that negatively affected the rating, Dmitry Medvedev named the 'low quality of state and local government, ineffective application of law, and our essential trouble called corruption, economic crime, excessive administrative, technical and information barriers as well as a very high level of monopoly' ('Russian president calls for improving investment climate', 2010).

Another factor that could have potentially impeded the Russia–Germany energy cooperation, but did not, is a long history of geopolitical rivalry between the two countries and the legacy of the Second World War. During the Cold War, the Germany–USSR relations had their ups and downs, with occasional periods of warming. The exports of Russian natural gas to Germany were growing steadily from the 1970s to 2010, from 1.1 billion cubic meters in 1973 to 39 billion cubic meters in 2010 (Stern, 2011). The fact that the two states could overcome the negative past of WWII has been quite remarkable, compared with other dyads of states with a history of mass killings (e.g. Armenia–Azerbaijan, Armenia–Turkey, etc.). For example, for the Baltic States, the wounds of WWII are still very fresh, and this is one of the reasons the relations between Russia and the Baltic States have been so tenuous. Russia and Germany, however, moved beyond those past casualties and have been continually strengthening their economic ties.

Third, Germany's decision to build this pipeline contradicted the overall energy policy of the European Union that aimed at diversifying energy supplies and reducing the imports of Russian natural gas. While leading the European Union in multiple issues (e.g. fiscal policy), Germany has taken a clearly separate path in ensuring its own energy security by increasing imports of Russia's natural gas and investing billions in new pipelines.

Finally, the Nord Stream pipeline was criticized and protested by environmental non-governmental organizations both in Europe and in Russia.

The environmentalists raised serious concerns about the pipeline causing possible damage to the Baltic Sea flora and fauna. Further complicating things, the Baltic Sea bottom is littered with chemical weapons, left there during WWII and since untouched. So far, the littoral states have preferred the chemical munitions to just sit there, and the concern was that the pipeline construction would create a disturbance that might cause a leakage of chemical munitions. Even though environmental groups are usually very influential in Germany on multiple issues (e.g. a nuclear-free economy, renewable energy, and others), in the case of Nord Stream, the environmental objections did not stop the implementation of the pipeline.

An analysis below explores all these factors and explains why they have not prevented Germany and Russia from close cooperation. On the contrary, the two countries put their differences aside and focused on their economic benefits, as well as geopolitical advantages on both sides. The governments of Russia and Germany were the major decision-makers in the Nord Stream pipeline planning, in line with their government energy policies. Germany has been interested in a stable supply of natural gas. Obviously, Russia's geographic proximity and rich natural gas resources make Russia an obvious supplier for Germany. However, these geographic factors alone are not enough to account for the increasing energy ties between the two states. The Nord Stream pipeline also adds to Germany's importance as a major energy hub and a reseller of natural gas in Europe, increasing the country's geopolitical importance.

Economic considerations for the Nord Stream pipeline

Russia's enormous resources of oil and natural gas predict high interest from the European partners in Russia's energy exports, in line with the economic explanation of energy cooperation. However, the investment climate in Russia for foreign companies that want to participate in oil and natural gas production has not been favorable, according to major rankings by international institutions such as the World Bank. In 2009–2010, when the Nord Stream pipeline was being implemented, the World Bank ranked Russia as 123rd out of more than 180 countries in terms of the Ease of Doing Business. Russia was ranked below Uganda in 122nd place ('World; World Bank: Russia Slips in Doing Business Rank Despite Reforms', 2010). However, at the time, the Medvedev Administration understood the problem and was working on administrative reforms. Despite the difficulties for foreign investors, German companies seemed to be operating in Russia almost seamlessly, which is partially explained by a strong economic rationale on both the Russian and German sides.

Russia's natural gas resources and exports to Germany

Russia is extremely rich in natural gas – sometimes the country is called the 'Saudi Arabia of natural gas' – as Russia possesses the biggest natural gas reserves in the world, similar to how Saudi Arabia holds the largest reserves of oil.

Similar to Azerbaijan's oil resource estimates explored in Chapter 3, energy experts have disagreed on the estimates of Russia's available reserves. Jonathan Stern, for example, pointed to the large discrepancies in various sources of data. The 2005 Russian Energy Strategy, issued by the Russian government, estimated Russia's natural gas reserves as 127 trillion cubic meters. Stern noticed, however, that this number was only half of the estimate provided by the Russian government just two years before that, in 2003 (Stern, 2005, p. 1).

Most of the natural gas production in Russia has been concentrated on the so-called 'giant' fields. The three most notable fields – Bovanenkovskiy, Harasaveiskiy, and Novopostovskiy (Yamal) – hold the combined reserves of 5.9 trillion cubic meters. Some giant fields have been developed by Russian government-controlled companies jointly with international ones. Some examples include the Shtokman field in the Barents Sea (3.7 trillion cubic meters of natural gas), developed jointly with Total (France) and Statoil (Norway); the South Russian field (Yamal, 1 trillion cubic meters), developed with Wintershall and Ruhrgas (Germany); and the Kovistinskiy field (East Siberia, 2 trillion cubic meters), developed with TNK-BP (Yazev, 2010, p. 76).

Germany, on the other hand, produces only 20 per cent of the natural gas it consumes. The remaining 80 per cent is imported from the following countries: Russia (39 per cent of the overall German gas imports), Norway (26 per cent), the Netherlands (23 per cent), and others (12 per cent) ('European Market', 2011). Natural gas has been delivered to Germany via a complicated system of pipelines. Importing gas from Russia, before Nord Stream, Germany had been heavily dependent on transit countries, such as Ukraine, Poland, and Belarus. The flow of natural gas from Russia to Western Europe has been regularly interrupted because of price disagreements between Russia and Ukraine. Two of the most disruptive disputes happened in January 2006 and in January 2009. Western Europe, including Germany, experienced a sharp fall in the amount of natural gas delivered via Ukraine during those disputes, which was especially frightening because it usually happened during the coldest month of the year (Bierman & Meyer, 2009).

Currently, about three quarters of Russian natural gas intended for the EU goes through Ukraine. Ukraine has long been paying a subsidized price for Russian gas. During the Russia–Ukraine natural gas disputes, Western European countries did not receive their full import volume of Russian gas, because Ukraine was using some of it for its domestic needs. During the disputes, Hungary and Poland received only half of their gas imports, while France, Slovakia, and Croatia received only one third. Even Germany and Italy, despite their close ties to Russia, complained about reductions in the natural gas flow. Ukraine denied any siphoning of gas intended for the European consumers, while Gazprom claimed that the company provided gas for Europe in full, deducting only what had been intended for Ukraine because of Ukraine's multi-billion debt to Russia for the previous supplies (Aalto, 2008a, p. 38). For the European Union, which imported 25 per cent of its natural gas from Russia, every day of the delay in the Russia–Ukraine price negotiations was almost deadly. The situation was complicated, especially in 2009, by Ukraine's desire to join NATO, which was unacceptable by Russia. In the meantime, the European countries felt as if they were 'hostages' in those gas conflicts (Stack & Rotella, 2009).

For two decades since the breakup of the Soviet Union, Russia was selling subsidized natural gas to Ukraine. Providing cheap energy for neighboring states was Russia's way to keep the economic ties with the Near Abroad, as well as to influence the post-Soviet states. However, as Ukraine, as well as other post-Soviet countries, was politically moving in the direction of joining NATO and away from Russia, Russia started to bring the natural gas prices for Ukraine closer to the European market level. By 2012, the price for Ukraine was already close to the European level ('Gazprom Opredelilsya s Tsenoi na Gas Dlya Evropy: Ona Budet Nizhe, Chem Dlya Urkainy', 2012).

A year later, in 2013, Russian President Vladimir Putin and Ukrainian President Viktor Yanukovych singed a deal that stipulated the price of 1,000 cubic meters of natural gas at $268.5, compared to $400 the year before. At the time, Ukraine was considering entering the Russia–Belarus–Kazakhstan Customs Union, having rejected an Association Agreement with the European Union ('Russia–Ukraine Agreements; Russia Agrees to Lower Gas Price for Ukraine By a Third, Customs Union Not Discussed at Presidents' Meeting', 2013). The situation between Russia and Ukraine changed drastically in 2014, when Ukraine changed the course and announced the plan to sign a trade agreement with the European Union. The events that unfolded after that (annexation of Crimea by Russia and the Eastern Ukraine conflict) are discussed in detail in Chapter 5.

Gazprom charged similar low prices to Belarus, in exchange to the Belarus government's support of Russia, even though the Russian state budget was underpaid $2.5 billion annually because of it (Mazneva, 2009). In January 2015, the country imported 2 billion cubic meters of natural gas from Russia for the price of $142 per thousand cubic meters, which is even cheaper than the price for Ukraine ('Statistics; Belarus Ups Oil Imports 5.1%, Petroleum Product Exports 35.5% in Jan', 2015). Russia is aware that many industries in Belarus would go bankrupt if the country pays the European price for oil and natural gas. As a result, a financial compromise was reached in 2009 in which Russia received limited access to the privatization of energy companies in Belarus, such as the Naftan–Polimir company (Gudkov, 2009).

The fluctuation of prices and the disputes that Russia had with the transit states have been a subject of concern for Germany. Moreover, in addition to gas exports, Russia's Gazprom has participated in Germany's natural gas distribution system. In 1993, Gazprom and Germany's Wintershall formed Wingas – a joint venture that have owned and operated 2,000 km of Germany's gas pipelines, as well as the Rehden underground gas storage facility – the largest in Europe – with the capacity of 4 billion cubic meters (Gazprom, 2011). This joint business activity has deepened the economic ties between the two countries, despite the unstable business climate in Russia.

Russia and Germany cooperated not only in the economic and energy sector, but also in security aspects, such as a joint diplomatic initiative in the territorial dispute of Transdniestria, in which Russia maintained about 500 troops. In 2011, Angela Merkel and Vladimir Putin (then Prime Minister) met to discuss a joint resolution, which analysts called the 'first concrete instance of Russia and Germany working jointly to dictate the terms of key European security issues' (Anon, 2011). Russia and Germany presented their initiative as a diplomatic move:

> The resolution would call for Transdniestria to receive representation in the Moldovan parliament. In exchange, Russia will consider allowing a peacekeeping or monitoring force from the European Union or the Organization for Security and Cooperation in Europe (OSCE) into Transdniestria to help the Russian military patrol the region.
>
> (Anon, 2011)

In addition to the energy sector, the governments of Germany and Russia have actively supported business deals outside of natural gas and oil. For example, German carmaker Opel was acquired by Russian companies Magna and Sberbank. Former Russian energy minister Igor Yusufov bought Germany's Wadan Yards. Germany's Siemens has participated in building wind turbines and trains for Russia, along with establishing several research and design centers. The German Energy Agency and the Russian–German Energy Agency helped facilitate cooperation between Siemens and Russia's Renova StroiGroup in energy efficiency projects. Russian companies RusHydro and Russian Technologies, jointly with Siemens, have produced 500 megawatt of wind energy in Russia, helping the Russian government achieve the goal of a 4.5-per cent share of renewable energy in the country's energy consumption by 2020 (Filatova, 2010).

Such business initiatives have been implemented with less government involvement. These business transactions have contributed to the larger interdependence between the two economies ('Mi Izdadim Germano-Rossiiski Uchebnik Istorii: Interview with Walter Jürgen Schmidt', 2009). Merkel said in 2010 that cooperation became 'more specific, … more intensive and less complicated' (Filatova, 2010).

Nord Stream and its challenges

The repeated natural gas price disputes between Russia and Ukraine led Germany to question the reliability of the existing transit routes for natural gas. As a solution, both Germany and Russia proposed a direct pipeline between the two states, that later received the name 'Nord Stream'. Gerhard König of the Russian–German joint venture Wingas summarized the rationale for the pipeline as: 'Europe needs ways of accessing supplies that are independent of transit countries, like the construction of the Baltic Sea pipeline' (Intellpuke, 2009). As a result, Nord Stream, a 55-bcm per year natural gas pipeline, was built on the bottom of the Baltic Sea in the exclusive economic zones of Denmark, Finland, and Sweden, with the onshore terminals in Germany and Russia ('Nord Stream AG: Project Information Document. Offshore pipeline through the Baltic Sea', 2006, p. 1). Originally started by only Germany and Russia, eventually the pipeline became a multinational project. The shareholders of the pipeline included BASF's subsidiary Wintershall (Germany), E.On (Germany), Gasunie (the Netherlands), and the largest shareholder – Gazprom (Russia) (51 per cent) (Topalov, 2009). The Russian government controls 50.002 per cent of Gazprom's shares, but foreign nationals are allowed to buy Gazprom's shares (Gazprom, 2011).

Nord Stream was presented as a part of the Trans-European Energy Network (TEN-E) that connects the United Kingdom through Germany and the Baltic Sea to Russia ('Nord Stream AG: Project Information Document. Offshore pipeline through the Baltic Sea', 2006, p. 1). TEN-E was originally designed by the participating governments as 'integral to the European Union's overall energy policy objectives, increasing competitiveness in the electricity and gas markets, reinforcing security of supply, and protecting the environment' ('Energy Infrastructure: What Do We Want to Achieve?', 2008).

After responding to the environmental impact assessment review, the Nord Stream pipeline was built in just 16 months for 7.4 billion Euro (about $10 billion). Already in November 2011, state leaders Angela Merkel and Dmitry Medvedev, prime ministers of France and the Netherlands, Francois Fillon and Mark Rutte, and the EU energy commissioner Günther Oettinger, attended Nord Stream's inauguration ceremony in the village of Lubmin, Germany ('Germany, Russia to Launch Nord Stream', 2011). The 760-mile Nord Stream became the world's longest underwater pipeline. The pipeline capacity is an equivalent of supplying 26 million households with heat and hot water per year (Steininger, 2011). At the full capacity, the pipeline will pay for itself in 10–16 years, if the price of natural gas for Germany remains at about $450 for 1,000 cubic meters.

Germany has pursued a pragmatic approach in Russia, focused on trade and investments. As a result, Germany has been one of Russia's main trading partners, with the overall trade volume of 76.5 billion Euro in 2013 alone, which was still 5 per cent lower than in 2012. In 2013, Russia imported goods and services from Germany for 36 billion Euro. In the same year, Russia exported 40 billion Euro worth of goods and services (mostly energy resources) to Germany. In 2013, trade with Russia comprised almost 4 per cent of Germany's overall foreign trade and was Germany's 11th most important trading partner in 2013 and the 10th in 2012 (Khoroshun, 2014).

One of the factors that helped close economic relations between Germany and Russia, and especially the Nord Stream pipeline implementation, has been friendship between Russian President Vladimir Putin and former German Chancellor Gerhard Schröder (in power from 1998 to 2005). During Schröder's term in government, the two leaders used to regularly meet, and Schröder even joined the board of Nord Stream. At the pipeline inaguration ceremony Vladimir Putin said:

> From the first steps under this project, we have been supported by our German friends, and, primarily, our good friend Gerhard Schröder. For all the fuss that surrounded this project, he saw the value of this project and made for himself a fundamental decision and stood by it consistently.
> ('Putin Thanks Schröder, All Nord Stream Participants', 2011).

Schröder's involvement helped legitimize Nord Stream in the eyes of the German population (Szabo, 2009, pp. 26–28).

However, Schröder has faced criticism from other European governments and even from German conservative parties for being very close to Putin. In reply to critics, he defended the Nord Stream project and even supported Russia's gas pricing policies toward Ukraine. He pointed out in 2009 that Russia had offered a price to Ukraine that was below the level for the European Union, and therefore Ukraine had few reasons to be unsatisfied, especially since Ukraine did everything possible to be a part of the EU and independent of Russia. As a solution for repeated Russia–Ukraine disputes, Schröder suggested that an international consortium of companies be created to manage the Ukrainian pipelines, since the country had been unable to maintain the pipelines properly. He emphasized that through the Balkan Sea pipeline, Germany could become independent of transit countries, eliminating middlemen ('Gazprom Says European Gas Deliveries Could Resume Friday', 2009).

Chancellor Angela Merkel, Schröder's successor, has mostly continued the pragmatic approach to Russia practiced by her predecessors, especially during the time when the Nord Stream pipeline was discussed and constructed. The government meetings were followed by business conventions for top-level politicians and executives from both countries, such as the 2010 Russian–German Commodities Forum in Freiberg, Germany. The participants included Prime Minister of Saxonia, Stanislaw Tillich; Deputy Federal Minister of Economics and Technology, Jochen Homann; Saxonia's Minister of Finance, Georg Unland; Minister of State of Foreign Ministry, Cornelia Piper; and, on the Russian side, President of Russian Gas Society, Valeriy Yazev (Bandlova, 2010, p. 10). The forum was established in 2006 by the major mining universities of Germany and Russia, under the aegis of both governments, in order to 'bring together manufacturers, politicians, and academics' (Filatova, 2010, p. 11) and to ensure an uninterrupted supply of raw materials from Russia to Germany. The Nord Stream pipeline was a regular topic discussed in such bilateral meetings. Major German government agencies have also been involved in bilateral financing of deals with Russia: among them, the Federal Ministry for Economic Cooperation and Development (BMZ), the Agency for Technical Cooperation (GTZ), KfW Development Bank, and the Foreign Trade and Investment Promotion Scheme (AGA) (Razavi, 2007, p. 134).

In Europe, however, not everyone was happy about the construction of the pipeline. For some, the pipeline was another milestone in a global 'battle' over energy resources. Alexander Rahr of the German Council on Foreign Relations in Berlin put it succinctly: 'The 21st century will be the century of fights over natural resources... We will see new alliances struck all over the globe attempting to create energy security' (Steininger, 2011). Critics voiced several concerns about whether Russia can produce enough natural gas in order to be able to meet Russia's export commitments to Western Europe. Although the country's natural gas reserves are indisputably large, the infrastructure has been lagging behind.

First, Russia's oil and gas production technology suffered from the lack of investments in the 1990s, which affected the current state of the industry's technological development. Russia's oil and natural gas production is more energy intensive and less efficient than that of developed countries. For example, the newest technology, such as the liquefied natural gas production (LNG), was introduced in Russia only recently, as part of the export diversification effort. Natural gas, compressed and cooled to the temperature of –260°F may be delivered around the world in liquid form by tankers, unlike the conventional natural gas that can only be delivered by pipelines in gas form. Currently, only one LNG plant is operational in Russia – Sakhalin-2, and the Russian government plans to build three other LNG plants –Yamal, Shtokman, and Baltic (Yazev, 2010). Another widespread technology, at least in the United States, is hydraulic fracturing (fracking), which has not been developed in Russia. According to Gazprom, shale gas will not be feasible in Russia for another decade, because of large deposits of conventional natural gas (44.6 trillion cubic meters as of 2013), which is much cheaper to extract (Robertson, 2013).

The second concern by the critics has been that geographically, most new natural gas production sites in Russia are located in East Siberia, the Russian Far East (including Sakhalin), and the Arctic. The Arctic is a region of increasing importance for all the Arctic states, and especially for Russia. However, this is the most technologically challenging and costly region to produce oil and gas due to the harsh weather conditions, half-a-year night conditions, ice, and lack of infrastructure. The environmental dangers of producing oil and gas here are the highest as well, as any accident or spill would be fatal for the fragile environment in the Arctic. In the Arctic region especially, Russia needs cooperation with foreign companies that have expertise in such conditions, and such cooperation might be easily affected by foreign sanctions and geopolitical disagreements. Nevertheless, some international companies have been allowed to work there even under the present political conditions: Total (France) received permission by the French government to participate in Arctic projects – Yamal LNG, Kharyaga and Termokarstovoye. Total (49 per cent of the project) and Russia's Novatek (51 per cent) will jointly produce up to 16.5 million tons of LNG per year just from Yamal LNG ('Gas; France Clears Total to Work at Yamal LNG, Kharyaga and Termokarstovoye', 2015).

To overcome potential shortages in Russia's natural gas production that might arise from the Western sanctions on equipment, experts suggested another way to meet Russia's export commitments to Europe: namely, to purchase natural gas from Turkmenistan and Kazakhstan and to deliver it using Russia's pipelines. However, this might be not feasible in the future, as the Central Asian states set their natural gas prices closer to the European level and try to bypass Russia via pipelines in the Caucasus: e.g. via the Trans-Caspian pipeline. The 300-kilometer, 30-billion-cubic-meter Trans-Caspian pipeline would connect Turkmenistan and Azerbaijan on the bottom of the Caspian Sea and would then continue to Europe. Turkmenistan has already started building its precursor – the East–West pipeline – that will deliver natural gas from all over Turkmenistan to the main terminal on the Caspian shore of Azerbaijan ('Turkmenistan believes in Trans-Caspian pipeline project', 2015).

Another potential solution for the Russian government in their fulfilment of contracts with Western Europe is to buy natural gas from Russia's smaller independent companies and resell it to Europe via Gazprom's monopolistic system of pipelines (Hanson, 2009, p. 42).

In Russia, critics of the pipeline also question its economic feasibility. Opponents of the pipeline, such as the former Russian deputy energy minister, Vladimir Milov, describe it, as well as other pipelines bypassing Ukraine and Poland, as a political project, expensive and problematic. Milov argues that it is cheaper and more efficient to transport natural gas by land via Ukraine. He further insists that Russia should more actively participate in international energy institutions and charters rather than trying to find a 'perfect' transit route. Milov, agreeing with environmental NGOs, also emphasizes possible environmental damage to the Baltic Sea. As a more economical option, he recommends that Russia should reach a more stable agreement with Ukraine and lay another gas pipeline via this country, parallel to the existing pipeline infrastructure (Grigoriev & Milov, 2009). Those recommendations were quite plausible back in 2009, however, after the events in Ukraine in 2014–2015 (explored in detail in Chapter 5), it is obvious that the normalization of Russia–Ukraine relations might now take decades.

In defense of Nord Stream, however, another prominent expert, Leonid Grigoriev, argues that international regimes are not effective enough to protect Russia's interests and to ensure that the transit countries comply with their contractual obligations. He refers to a series of disputes between Russia and Ukraine over transit fees, gas prices, and Ukraine's debt to Gazprom. Those episodes sent chills across Western Europe, as they literally happened in the middle of winter. Grigoriev concluded that using transit countries was risky, both for Russia and for Germany. Furthermore, he concludes, Nord Stream was started as a German initiative and a profitable commercial project (Grigoriev & Milov, 2009).

Nord Stream did not stop Germany's policy of natural gas supply diversification: the country has been establishing economic ties with as many energy exporters as possible. Nord Stream has been just one element in Germany's natural gas supply system. The pipeline provides Germany with exceptional access to Russian gas, increasing Germany's importance in the European energy redistribution system (Szabo, 2009, p. 29). In addition to using natural gas for its domestic needs, Germany has been focused on creating an extended system of natural gas storage, which currently has 46 natural gas facilities in order to secure gas supplies in case of import shortages. The Germany Energy and Water Association's chief Martin

Weyand estimated in 2009 that the facilities stored enough gas for 40 days of consumption in case of interruptions in imports ('Europe Pushes for Russia to Resume Gas Deliveries', 2009).

In addition to the diversification of exporters (Russia, the Middle East) and transit routes (Ukraine, Poland, Belarus, the Baltic Sea route), Germany has also worked on the diversification of the country's energy mix (natural gas, oil, nuclear, renewable energy) ('Europe Pushes for Russia to Resume Gas Deliveries', 2009). This mix has been reviewed after the Fukushima nuclear disaster in 2011, when Germany announced its plan to phase out nuclear energy by 2022. Currently, the German nuclear energy operators Vattenfall, E.ON, EnBW, and RWE are working on the phase-out plan, which is estimated to cost about 70 billion Euro ('Costs for Germany's Nuclear Exit Might Increase to 70 Billion Euros', 2015). The 18 nuclear power plants in Germany currently produce about 28 per cent of the country's electricity (Daly, 2011). The nuclear exit will potentially increase the need for natural gas as a transition fuel.

Simultaneously with phasing out nuclear energy, Germany has been working on reducing greenhouse gas emissions and the development of renewable energy. Some of such projects have been done jointly with Russian counterparts. For example, Germany's RWE partnered with Russia's Kurchatov Institute and the Skolkovo Foundation to innovate in the areas of renewable energy generation, such as biomass, and energy efficiency ('RWE, Russians Explore Energy Efficiency', 2011).

The European policy of diversification

Germany's trend to diversify energy supplies has not been unique in the EU. The European Union as an institution, too, has prioritized diversification over the last decade. A major goal in this policy has been to reduce dependence on Russian natural gas. The percentage of Russian natural gas in domestic natural consumption is 39 per cent for Germany, 64 per cent for Hungary, 69 per cent for Austria, 77 per cent for Czech Republic, 82 per cent for Greece, 98 per cent for Finland, 99 per cent for Slovakia and Bulgaria, and 100 per cent for Estonia, Latvia, Lithuania, and Moldova. In other countries, such as France, Italy, Poland, Romania, and Slovenia, the percentage of Russian natural gas is from 22 to 43 per cent – still a substantial amount. Even the Netherlands, a natural gas producer itself, has 6 per cent of its domestic consumption imported from Russia. Overall, the European dependency on Russia is even higher, as some natural gas from Turkmenistan also goes to Europe through Russian pipelines (Gelb, 2007, p. 1).

The diversification of energy supplies in Europe is not a recent phenomenon. Already as early as in 1981, studies such as 'The European Transition from Oil' explored several alternatives to oil dependency, such as coal (which even in 2008 still provided one third of EU energy needs), nuclear energy, wind, solar power, and natural gas. In those years, however, the USSR was considered the main supplier of Germany's natural gas, as well as oil and uranium, and the country's reliability as an exporter was not put into question (Aalto, 2008a, p. 24).

As a result of the EU diversification efforts, the share of Russian gas in European overall natural gas imports decreased from 38 per cent in 2009 (Papava et al., 2009, p. 26) to 30 per cent in 2014. Other countries supplying natural gas to the EU are Norway, Algeria, Nigeria, Libya, and Egypt. Considering the unstable situation in Libya and Egypt for the last several years, these two countries are unlikely to offer required stability. In addition to that, natural gas supplies from the Middle East and Africa have been delivered mostly in the LNG form, which is more expensive than natural gas delivered by conventional pipelines ('Wall Street Sector Selector: Liquid Natural Gas Europe's Answer to Dependency on Russia', 2014). Nevertheless, the EU states managed to substantially reduce the share of their combined gas imports from Russia from 66 per cent in 1990 to 48 per cent in 2007 (Papava et al., 2009, p. 37). It was done, for example, by increasing gas imports from Norway by 27 per cent and from Qatar by 115 per cent in 2009 (Grib, 2009).

The instability in the Middle East caused some states to increase their reliance on Russia. Because of violence and political instability in the Middle East and North Africa, Italy has actually increased its share of Russian natural gas, from 32 per cent in 2012 to 49 per cent in 2014. Other sources of natural gas for Italy, such as LNG from Algeria, might be readily available but more costly ('Italy/Russia: Italy Faces Dire Situation if Russian Gas Supplies Cut', 2014). The Russian government understands this well and uses this to its own advantage: Russia's Minister of Energy Igor Sechin stated in 2009 that it would be expensive for Europe to substitute the Russian sources (Zygar & Grib, 2009). Other options, such as shifting back to coal and oil, however, would be disastrous in terms of climate change effect.

Another potential source of natural gas for Europe is shale gas, although this source is very environmentally problematic. European states have to decide between the need to increase their domestic natural gas production and the environmental degradation that such production might cause. For example, Poland discovered shale gas reserves that might potentially increase their energy security. In 2011, the U.S. Energy Information Administration estimated Poland possessed 5.2 trillion cubic meters of shale gas, which is an equivalent of a 300-year domestic supply. In 2012, according to Marcin Zieba, general manager of the Polish Exploration and Production Industry Organization, multiple companies (including Chevron, 3Legs Resources PLC, San Leon Energy PLC, and others) were 'working in accordance with their plans, testing reserves and looking with optimism to the outlook of future shale gas production' (Kruk, 2012).

Other companies were skeptical, however: ExxonMobil conducted test drilling in central Poland and announced that gas wells 'failed to yield commercial quantities, raising doubts about Poland's hope to replicate the U.S. shale gas bonanza and free itself from dependence on Russian gas imports' (Kruk, 2012). To encourage investors, in 2014, the Polish government implemented tax breaks for shale gas production. However, as of 2015, experts do not predict any rapid growth of shale gas production in Poland because the drilling results in October 2014 showed 'disappointing flow-rates' in the trial wells. Also, the number of wells drilled in Poland from 2010 to 2014 was only 13 wells annually, which was not enough even to measure the 'commerciality' of hydraulic fracking in the country ('Poland Oil & Gas Report – Q3 2015', 2015).

Poland has not been the only country experimenting with shale gas production in Europe. Bulgaria, too, has hoped that shale reserves would help it revisit its energy ties with Russia. In 2011, hoping to develop large-scale shale gas reserves, Bulgaria withdrew in 2011 from a large Russia-led pipeline project – the Burgas–Alexandroupolis oil pipeline – and was reconsidering its participation in the Russia-backed South Stream gas pipeline (later cancelled by Russia, anyway) (Assenova, 2011). However, in 2012, Bulgaria banned fracking for shale oil and gas production, because of environmental concerns, so as of 2015, this option for Bulgaria is not even on the table ('Bulgaria Oil & Gas Report – Q2 2015', 2015).

The environmental concerns about fracking (such as the risk of ground water contamination, methane emissions, waste water treatment, earthquakes, and others) have been very prominent in the European Union. In May 2015, following the example of Bulgaria and others, the German Parliament, too, discussed a possible ban on fracking, or at least strict limitations on where fracking can be done. In the same month, Denmark discontinued the first hydraulic fracturing operation, because Total (France), which was conducting drilling, was using an 'unauthorized chemical' in their fracturing fluid. So far, only the UK, Romania, and Poland have been fracking-friendly countries in the European Union ('Reluctant Minister Introduces Fracking Bill', 2015). Lithuania has also been considering the deployment of fracking technologies ('Western Europe Oil and Gas Insight – August 2014', 2014).

Because of all these developments, the task of diversification for Europe has not been easy. Russia's geographic position and natural gas reserves naturally ensure a very favorable position for Russia as the top natural gas exporter in the world, as well as a transit state for Kazakhstan, Uzbekistan, and Turkmenistan.

The policy of diversification is especially urgent during the Western sanctions on Russia, imposed in 2014. Previously, during the U.S.–Russia 'reset', and consequently warmer relations between Western Europe and Russia, the European Union considered the closer integration of Russia in the European economic space, as addressed by the European Commission in their Strategic Energy Reviews in 2010 (Yazev, 2010, p. 81). However, the situation with Ukraine in 2014 and the Western sanctions drastically changed those plans of integration.

In response to the new political context, Europe has been building new pipelines in order to deliver natural gas from multiple producers (including Central Asia and the Caspian). In addition to the implemented Nord Stream and the Baku–Tbilisi–Ceyhan (BTC) pipeline, the European Union was considering two other pipelines – South Stream (Russia-backed), and Nabucco (bypassing Russia). South Stream, a planned 26-billion-Euro pipeline with a maximum capacity of 63 billion cubic meters per year, would bring natural gas from Russia on the bottom of the Black Sea to Bulgaria, Greece, Italy, and Austria. Its rival

– Nabucco, a 15-billion-Euro natural gas pipeline with a maximum capacity of 10 billion cubic meters per year would go from Turkey via Bulgaria, Romania, and Hungary to Austria. In response to some critics who argue that Europe does not need that many pipelines, Germany instead advocated its own 'policy of multiple pipelines' that would all contribute to European energy security ('Mi Izdadim Germano-Rossiiski Uchebnik Istorii: Interview with Walter Jürgen Schmidt', 2009). German officials and executives, such as Gerhard Schröder, a Nord Stream and TNK-BP board member, and Klaus Mangold, the chairman of the Committee on Eastern European Economic Relations of the Federation of German Industry, often emphasize this policy as a way to achieve better energy diversification ('Europe Fails to Wean Itself Off Russian Gas', 2009). Either way, the South Stream was cancelled by Russia ('Russia Scraps South Stream Natural Gas Pipeline Project', 2014), and Nabucco was transformed into the Southern Gas Corridor (Kanter, 2015).

In 2015, the Southern Gas Corridor has been the most discussed pipeline, especially in light of possible close relations between Russia and Greece. The U.S. government has been trying to persuade the Greek government to join the Southern Gas Corridor, supported by the West, rather than other projects to Turkey, such as the Turkish Stream, pushed by Russia. The U.S. State Department envoy, Amos J. Hochstein, said that the Turkish Stream 'is not an economic project' but is 'only about politics'. The Southern Gas Corridor, he added, would be an 'excellent project for Greece as it will create a significant amount of jobs'. In response, Gazprom's CEO Alexei Miller asserted that the Turkish Stream was a 'real blessing for the entire European gas market' (Kanter, 2015). As of May 2015, the heated debates about the two pipelines continue in Europe.

The discussions about alternative pipelines and import volume have happened in the overall context of the EU–Russia relations. On the issues of energy supplies, the Russian government has preferred to deal with European countries on a bilateral basis (Klitsounova, 2005, p. 39). This can be partly explained by Russia's deliberate policies and partly by internal divisions on energy security among EU members themselves. However, since the mid-1990s, Russia and the EU have worked (not always successfully) on several treaties and agreements to institutionalize their relations, alongside European internal agreements regarding Russia.

The first legal framework of the EU–Russia cooperation was established in 1997 via the Partnership and Cooperation Agreement (PCA) that covered the energy sector, trade, capital movement, telecommunications, and transport (Johnson & Robinson, 2005, p. 6). Two years later, in 1999, the European Union adopted the Common Strategy on Russia that was created for the following goals: to ensure democracy promotion and rule of law in Russia; to support a political and economic transformation in Russia; and to share European experience in building modern political, economic, social and administrative structures. The strategy, however, was more of a rhetorical instrument rather than a plan for specific actions.

In the 1990s, the European Union provided Russia with significant economic assistance and investments and loans on favorable terms. One such program – the Technical Assistance to the Commonwealth of Independent States (TACIS) – was administered by the European Commission and assisted Russia's transition to market economy, regularly evaluating Russia's performance along long-term and immediate goals (McCann, 2005, p. 205). Under an agreement with Gazprom, TACIS provided 2.3 billion Euro to Russia in order to expand natural gas pipeline networks, therefore ensuring a stable flow of Russian natural gas to Europe (McCann, 2005, pp. 280–209).

In 1991, the EU introduced the Energy Charter Treaty (ECT), which provides regulations for the energy trade liberalization (Aalto, 2008b, p. 11). In 1994, Russia signed the charter but did not ratify it. The charter aims at the protection of investment, trade in energy materials and products, transit and dispute settlement. One of the key provisions is to make all parties combat market distortions and barriers to competition in economic activities in the energy sector. This provision opens individual energy markets to more foreign competition, and this is one of the reasons Russia has not ratified the ECT yet, even though the negotiations about it have been going on for more than a decade. Powerful energy market players (e.g., Gazprom) have opposed the treaty, so that they can keep their near-monopoly (Aalto, 2008b, p. 12). In 2009, Prime Minister Vladimir Putin announced that Russia had left the list of signatories of the ECT, completely abandoning the treaty (Grigoriev & Milov, 2009).

Despite disagreements on the ECT, in 2000, Russian Prime Minister Victor Khristenko and European Commission Director-General Francois Lamoureux started another initiative, the EU–Russia Energy Dialogue, designed to create an improved diplomatic and institutional framework for European trade and investment in Russia (Aalto, 2008b, p. 13). The dialogue's goals were to create market conditions and to facilitate reforms: to reduce Russian natural resource monopolies, to improve Russia's investment climate and the access by foreign companies to Russia's resource extraction; to improve energy efficiency; and to develop a free market system, competitiveness and openness in the Russian energy sector. These goals were designed in order to ensure a stable energy supply to Europe and to reduce the chances of Russia's geopolitical manipulation (Morozov, 2008, p. 48). The end result of the dialogue's implementation would be a reduced dependency of Europe on Russia (Aalto, 2008b, p. 27).

For over a decade since its start, the dialogue served as a platform for several conferences and initiatives, as well as a place for dispute resolution during such events as the Russia–Ukraine gas dispute of 2009. On February 24, 2011, the dialogue's framework was expanded when EU Commissioner Günther Oettinger and Russian Energy Minister Sergey Shmatko signed three additional documents: the EU–Russia Early Warning Mechanism in the field of energy, a Joint Statement to Create a Joint Gas Advisory Council and a Common Understanding on the Preparation of the Roadmap of the EU–Russia Energy Cooperation until 2050. The EU–Russia Energy Dialogue has been accepted by Russia much more favorably than the Energy Charter because Russia can see in the dialogue the value for its energy trade with Europe and protection of Russian interests during disputes. It is different from the Energy Charter: Russian officials have pointed out that even in Europe itself, states protect their own national energy companies, despite the charter's requirement for trade liberalization (Lyutskanov, Alieva, & Serafimova, 2013).

An attempt to negotiate with the European Union as a unified body happened in May 2009, during an energy partnership council between the EU and Russia. Russian Vice-Prime Minister Igor Sechin negotiated with European Commissioner Andris Piebalgs about the Russia–Ukraine natural gas crisis of January 2009 and the upcoming oil and gas supplies from Russia to Europe. Sechin insisted that the European Union help finance gas storage on the Ukrainian territory for further transporting to Europe (Zygar & Grib, 2009). The two officials also discussed the EU Energy Charter: Russia offered some changes to the charter, but they were not accepted because the treaty had already been ratified by the EU members (Zygar & Grib, 2009).

An issue that has stirred fears in Western energy consumers has been Russia's initiative to organize natural gas exporters in the Forum of Gas Exporting Countries, which is often compared to the Organization of the Petroleum Exporting Countries (OPEC). Some Western analysts predict that the creation of such a forum may lead to tightly controlled and manipulated natural gas markets. The forum, established in 2001 in Iran and officially registered in the United Nations, currently includes the following members: Russia, Qatar, Iran, Algeria, Bolivia, Venezuela, Egypt, Libya, Equatorial Guinea, Trinidad and Tobago, Norway (an observer), and the Netherlands (an observer). In 2009, the unanimous election of Russia's Leonid Bokhanosky as Secretary General was a sign of Russia's leading position on the organization. Although the Russian energy sector's decision-makers denied that the forum would be similar to OPEC, they expressed hope that the forum would become a working mechanism to protect its members' interests. Russian Energy Minister Sergey Shmatko underscored that the forum would be especially important for future coordination in the world LNG market ('Rossiya Ne Dolzhna Upustit' Svoi Shans [Russia Should Not Miss Its Chance]', 2009, p. 41). In an attempt to further formalize Russia's role as the leader in natural gas issues, President of the Russian Gas Society, Yazev, has proposed to create the world energy code, as a new legal basis for international energy cooperation (Starikov, 2010, p. 7).

In 2011, in Doha, Qatar, the Gas Exporting Countries Forum convened for its first summit. In 2013, Russia hosted the second summit of the organization, and the members at the time included Algeria, Bolivia, Egypt, Equatorial Guinea, Iran, Libya, Nigeria, Oman, Qatar, Russia, Trinidad and Tobago, the United Arab Emirates, and Venezuela, with Norway, Iraq, the Netherlands, and Kazakhstan being observers ('Russia to Host Gas Exporting Countries Forum Summit July 1–2', 2013). As of 2014, the Forum conducted negotiations with other gas-producing countries. For example, in November 2014, Secretary General of the Forum, Seyyed Mohammad Hoseyn Adeli, visited Azerbaijan to discuss possible ways of cooperation ('Gas Exporting Countries Forum Seeks Cooperation with Azerbaijan', 2014).

The creation of the Gas Exporting Countries Forum only emphasized the mutual dependency between Russia and Europe in terms of natural gas trade. While the Western rhetoric usually focuses on Europe's dependency on Russia, Russian analysts, on the contrary, usually emphasize Russia's vulnerability and dependence on a large consumer (i.e. the European Union). Moreover, Russian experts often view such mutual dependency between the producer and the consumer as a relatively positive development. Such views were especially prevalent in Russia about five years ago, during the warming of the relationships between Russia and the West, and several Western experts also agreed that such mutual dependency existed: Russia needs European financial, technical, and managerial resources, while the EU needs Russia's energy resources, stable trade conditions and favorable investment terms (Green, 2010).

In his interview to Deutsche Welle in 2010, Leonid Grigoriev of Russia's Institute of Energy and Finance was very optimistic about Russia–Germany and Russia–EU energy cooperation. He stated that the relationships between Russia and Europe in the energy sector were 'warm'. As Russia's energy infrastructure, mostly built during the Soviet period, was to a large extent designed to supply Europe with oil and gas, the relations between Russia and Europe were benefiting from that set-up. Even though Russia was dependent on Europe for oil, coal, and gas revenues, he added, such dependency was mutually profitable. To continue that lucrative cooperation, according to Grigoriev, both Russia and Europe would benefit from expanding the energy infrastructure (i.e. underground gas storage and gas pipelines). The marine pipelines (on the bottom of the Baltic Sea and the Black Sea), especially, were to eliminate costly intermediaries and transit fees, therefore reducing energy conflicts. Grigoriev concluded that Russia and Europe needed to synchronize their energy policies and energy plans (Gurkov, 2010).

At the time, Western experts, such as Adnan Vatansever of the Carnegie Endowment, were equally optimistic about the prospects of Russia–Europe energy cooperation. He noticed that the access for foreign companies to Russia's oil and gas reserve exploration was being broadened, as Russian companies did not have enough capital and expertise. Even the reduced gas demand and lower prices in Europe because of the global recession of 2009 were favorable for Europe, as the feeling of energy insecurity by European countries was diminished (Vatansever, 2010). In 2013, Vatansever saw the change in the business climate, as Russia and the European Union were more and more 'at odds on the price of gas following the launch of the antitrust investigation against Gazprom by the European Commission' (Koranyi & Vatansever, 2013).

Russian analysts usually point to internal disagreements in the European Union as a reason for Russia's lack of progress in dealing with the EU as an international body. Vagif Guseinov, Director at Russia's Institute of Strategic Studies and Analysis and Editor-in-Chief of Vestnik Analitika, argued that Russia and the EU could find common ground in the energy sector if Russia had a stake in the European energy networks and Europe had a stake in exploration projects in Russia. In the meantime, he continued, Europe accused Russia of being a non-reliable partner (Guseinov, 2008, p. 432). He stressed that the internal disagreements in the EU prevented a constructive dialogue with Russia, because each EU member had the power to block the whole decision-making process (Guseinov, 2008, p. 430). Russia's Minister of Energy Sergey Shmatko, noted in 2009 that since Germany and Italy were cooperating with Russia, the approval of the EU as a whole was not as vital (Zygar & Grib, 2009). As a result, since the 1990s, there have been few attempts to negotiate with integrated Europe as a body, because of the disagreements between 'new' and 'old' EU members and the lack of a common policy.

Guseinov also pointed out that the EU had insisted that Russia sign the European Energy Charter, which was designed to promote free trade, but simultaneously built barriers for Russia's investments in the European energy sector. Moreover, in some countries, he added, such as Estonia, the Russian energy businesses were being squeezed out of the transit business. Russia's loss of influence in the Baltic states could result for Russia in the loss of cheap transportation routes and the loss of the near-monopolistic position in energy supplies (Guseinov, 2008, p. 455).

Several Russian experts disagreed with the mainstream Russian view, summarized above by Grigoriev and Guseinov. One alternative voice in Russian politics is Vladimir Milov, a former Russian Deputy Energy Minister (2002) and the president of the Energy Policy Institute (IEP). Milov has been a vocal critic of mainstream decisions by the Russian government on such projects as the Nord Stream pipeline, and deals with Ukraine, etc. In the late 2000s, the institute has been restructured as a private consultant company instead of an NGO, due to 'adoption of the new restrictive NGO legislation, that had substantially complicated the operations of NGOs in Russia' ('Think Tank: Institute of Energy Policy (IEP)', 2010).

Milov disagreed with the decision by the Russian government to leave the Energy Charter Treaty in 2009, as he believed that the charter would have been beneficial for Russia's energy trade. According to Milov, if Russia ratified the charter, it could keep the same contracts with its current European partners, but it would be easier to sign new deals. Also, it would be easier to transport Central Asian gas and receiving transit fees, rather than re-selling of Turkmen gas, which is not too profitable for Russia. Milov added that Russia's contract conditions, such as the separation of transit fee and product price, were already close to the charter's requirements. He disagreed with the official government position that stated that Russia did not need to be a part of established international institutions, such as the charter, which is de-facto the 'energy WTO' (Grigoriev & Milov, 2009).

Milov's colleague, Leonid Grigoriev, former Minister of Energy and President of the Institute of Energy and Finance, agreed with Milov that the Energy Charter was not completely in line with the other European laws and principles. He added, in defense of the charter, that the Ukrainian gas agreements with Russia were in fact complying with the charter, separating the natural gas price and transit fees. Grigoriev also noted that the charter was a 'gas document' because it covered mostly the issues of the natural gas trade (Grigoriev & Milov, 2009).

In general, Russian experts agree that the diversification of European natural gas supplies in the long run might even be beneficial for Russia. First, it would encourage the Asian export routes. Second, Europe would be buying Russian oil and gas without fear of over-dependency on Russia. Third, it would help Russia to diversify its economy, shifting away from hydrocarbon exports domination (Morozov, 2008). Russia has had its own strategy toward the EU with a broad agenda, beyond oil and natural gas (Johnson & Robinson, 2005, pp. 7–8). One idea was to have the Common European Economic Space (CEES) between Russia and the European Union (Johnson & Robinson, 2005, p. 15). Russia has seen itself as an 'energy bridge' between the West and the East, not only as a producer and an exporter, but also as a logistics hub in the East–West transport corridor of global energy resources. This role is supposed to be achieved via pipelines and the development of LNG markets. However, Russia's LNG production in the Arctic and in the Sakhalin Island has been developing rather slowly ('Gazprom Losing Battle for World LNG Market', 2010). In the meantime, major energy companies (ExxonMobil, Chevron, Royal Dutch Shell) keep signing long-term LNG supply contracts with China, India, Australia, and South-East Asia ('Gazprom Losing Battle for World LNG Market', 2010).

Geopolitical rivalries

As discussed above, in the economic domain, Germany's behavior in energy security seems to contradict the overall EU energy policy, i.e. the diversification of energy suppliers and reducing Russia's dominance of the European and global natural gas markets, as well as Russia's influence in the Near Abroad and in the West. Germany's divergence from the EU position is puzzling, considering that Germany is one of the founding states for the European Union and currently one of the pillars of the whole European economy.

The next aspect that deserves close attention is how Germany and Russia have managed to overcome the geopolitical rivalries and historical enmities of WWII and the Cold War. Germany's position on Russia has differed from that of many other EU members not only in energy security but also in such geopolitical aspects as NATO's expansion, the Iraq War of 2003, Georgia, Ukraine, and others. Germany and Russia have been able to put their enmity aside, partly because of the affinity and friendship between the two countries' leaders, a long history of economic ties and import–export operations that took place even in the Cold War, and Germany's divergence from other NATO members' positions on key geopolitical issues of the last two decades.

The Cold War period and beyond

The Russia–Germany relations, and, previously, the USSR–Germany ties, have experienced ups and downs over the course of post-WWII history. The Second World War itself left deep scars in relations between the two countries.

During the Soviet time, when Germany was divided, the Soviet government portrayed West Germany and East Germany in opposite ways. West Germans were usually portrayed as the successors to the Nazis, while East Germans were presented as the 'good' Germans who had almost nothing to do with fascism during WWII (Stent, 2000). In relation to West Germany, historians agree that 'German–Russian relations have often been extremely tense and militarized. Despite some periods of cooperation, the two states have rarely been formally allied and have more often been members of opposing blocs' (Newnham, 2002).

Shortly after WWII, for about a decade, Germany and Russia barely had any diplomatic relations. Not surprisingly, after the atrocities of WWII, resentment and distrust prevailed between the two countries in the after-war period, especially during the Soviet-imposed blockade of Berlin in 1948–1949. In 1953, the USSR suppressed an uprising in the German Democratic Republic (GDR), which further halted negotiations between the USSR and West Germany. The first trade agreement between the two countries was signed only in the late 1950s (Newnham, 2002, p. 110).

In 1955, German Chancellor Konrad Adenauer visited Moscow for the Adenauer-Khrushchev summit, signifying the importance of Germany as an economic partner for the Soviet Union and other countries of the Soviet bloc. In the late 1950s, trade between the two countries increased substantially from DM 262.9 million in 1955 to DM 689.6 million in 1958. However, in 1962–1963, the German government used embargoes on grain and large-diameter pipes, trying to make the USSR dismantle the Berlin Wall (Newnham, 2002).

Until the late 1960s, the USSR–Germany relations were mostly confrontational in the political realm. Nevertheless, since the late 1950s, the Soviet government already established economic ties with the Federal Republic of Germany (FRG). In 1969, the Soviet leader Leonid Brezhnev started the process of rapprochement with West Germany, which continued during the 1970s, and established contacts with the leader of the Social Democratic Party of Germany Willy Brandt. Simultaneously, the Soviet Union was de-facto controlling the East German regime by the so-called Brezhnev Doctrine that stipulated that the East European countries were supposed to follow the Soviet way of life. The USSR enforced its ideological and economic model in East Germany (Stent, 2000).

At the same time, in the 1960s, FRG started a policy of *Ostpolitik* toward the Soviet Union under the leadership of Willy Brandt, Chancellor of West Germany from 1969 to 1974 and previously Chairman of the Social Democratic Party. This policy was designed to improve relationships with the Eastern Block, and especially with the USSR. Similar to Mikhail Gorbachev two decades later, Brandt promoted dialogue and open exchange of ideas, winning the 1971 Nobel Peace Prize for his policy (Wiegrefe, 2010).

Both during the icy relations pre-1960s and the détente by Willy Brandt in the 1970s – over several decades from 1945 to 1987 – the USSR continuously imported German manufactured goods and exported raw materials: because of its technological backwardness in some consumer goods areas, the USSR was more dependent on Germany than Germany was dependent on the USSR (Newnham,

2002, p. 109). In the 1970s, Germany negotiated several aspects with the Soviet Union, 'from travel rights for citizens of West Berlin to preserving German unification as a possible future legal option' (Newnham, 2002, p. 40). The issue of German unification was always at the center of German policy concerns and considerations.

When Mikhail Gorbachev came to power in the 1980s and started the policy of *perestroika* and *glasnost* in the Soviet Union, he also relaxed the political and military grip on the East European bloc. The so-called Sinatra Doctrine, jokingly introduced by Gorbachev, was the policy of allowing East European states to decide their way of life for themselves. Gorbachev made relations with the European Union one of the biggest priorities in his foreign policy, especially after the collapse of Soviet-controlled space in 1989 and Germany's unification. Inside the Soviet Union, however, there was a lot of opposition about unified Germany's NATO membership. The conservatives inside the USSR blamed Gorbachev for 'giving GDR away'. The early relationship between Chancellor Helmut Kohl (in power from 1982 to 1998) and Mikhail Gorbachev were icy, and Kohl was even quoted as comparing Gorbachev to the Nazi leader Josef Goebbels, although later Kohl denied saying that (Miller, 1992).

East and West Germany were officially reunited on October 3, 1990, about a year after the fall of the Berlin Wall on November 9, 1989. On December 25, 1991, the Soviet Union was officially dissolved. The key event that facilitated the reunification was Gorbachev and Kohl's meeting in Russia on July 14, 1990. Germany's membership in NATO was a precondition negotiated before the meeting. That meeting was when their personal relationship became warm and friendly, which was very important 'in a period of rapid change like the end of the Cold War', as experts, including Condoleezza Rice, emphasized after the events. Germany agreed to cover the cost of relocating the Soviet troops stationed in Germany back to the Soviet Union (about 40 billion Euro) (Hellfeld & Chase, 2010).

As Gorbachev initiated the Soviet Union's transformation, opened the country to the Western values and made substantial concessions on the issue of German reunification and NATO's expansion, the relationship with unified Germany rapidly improved. The dissolution of the USSR in 1991 brought hopes of democratization and liberalization of the newly independent Russia (Buerrieri, 2012). After the breakup of the Soviet Union and the German reunification, the European Union, and especially Germany, provided strong economic and political support to Russia (Stent, 2007, p. 417).

Since then, the leaders of Germany and Russia have had warm relations and even personal friendship: examples include Chancellor Helmut Kohl and President Mikhail Gorbachev; Helmut Kohl and President Boris Yeltsin; Gerhard Schröder and Vladimir Putin; and Angela Merkel and Dmitry Medvedev. The current German incumbent Angela Merkel started her political career in East Germany before the German unification, and she participated in political and social transformations in her country. However, German commentators noticed that Merkel's relations with Vladimir Putin lacked warmth and friendliness and instead were business-like and pragmatic: 'Not like in Schröder's time – At their meeting in Dresden, Merkel and Putin looked cool in their attitude towards each other and uninspired' ('German Commentary Finds Putin, Merkel "Cool" Towards Each Other', 2006).

This lack of personal friendship between Merkel and Putin has been further complicated by the Western criticism that Russia has been using its energy resources as a foreign policy and even an 'energy weapon'.

'Yesterday tanks, today oil'

Even though the European Commission officially approved Nord Stream in 2000, the critics of the pipeline were very vocal. As the New York Times wrote at the beginning of the pipeline's construction, Gazprom, via Nord Stream, was 'driving a political wedge between Eastern and Western Europe', in order to 'lead to a new era of gas-leveraged Russian domination of the former Soviet bloc'. The Western commentators continued that '[w]ith its gas wealth and eyebrow-raising network of personal ties, Russia has divided members of the European Union that have vowed to act collectively to protect their security' (Kramer, 2009).

Eastern European countries had made multiple attempts to block the Nord Stream pipeline project, and the Baltic States were especially vocal opponents. Latvia, for example, tried to lobby the Russian government against the construction of Nord Stream, targeting mostly the Liberal Democratic Party of Russia (LDPR) politicians in the Russian Parliament (Duma) (Tolstikh, 2006, pp. 85–86). Latvia also tried to secure the support of the U.S. government.

Germany and Russia have made multiple attempts to discuss and evaluate historical contentious events, such as the Molotov–Ribbentrop Pact and the Katyn massacre. As the German Ambassador to Russia, Walter Jürgen Schmidt put it, Germany's approach is to admit the mistakes of the past and to find ways not to repeat them. The German government even suggested in 2009 that a German–Russia history textbook should be written by historians from both countries, in order to reach a balanced interpretation of the historical past and to alleviate fears of history falsification ('Mi Izdadim Germano-Rossiiski Uchebnik Istorii: Interview with Walter Jürgen Schmidt', 2009).

The Nord Stream pipeline was even compared by the critics in the West to the 1939 Molotov-Ribbentrop Pact: '…the pipeline issue evokes deep memories of a darker era of occupation and collaboration, and has become a proxy debate over Russia's intentions toward the lands it ruled from the end of World War II to the fall of the Berlin Wall' (Kramer, 2009). Zbigniew Siemiatkowski of Poland has summarized Russia's energy export to Europe as an energy weapon: 'Yesterday tanks, today oil' (Kramer, 2009).

During the discussions of the Nord Stream pipeline, Latvia and the other Baltic States repeatedly connected the pipeline issue to the events of WWII. In 2009, Nils Muiznieks, Director of the Advanced Social and Political Research Institute at the University of Latvia, during his talk in the United States, emphasized that the Baltic States and Poland were representing a united front and the 'anti-Russian axis', in order to reduce the dependency on Russia for energy resources and electricity (Muiznieks, 2009). Muiznieks urged the European Court of Human Rights to issue a ruling that the Baltic States were victims of Soviet aggression during WWII, and that Russia should be forced to pay compensation for the occupation. Muiznieks further stressed that only integration with the European Union and NATO helped the Baltic States to avoid Russian domination after the end of the Cold War (Muiznieks, 2009). Negotiations between the Baltic States and Russia have been very difficult because of those historical enmities. However, the Baltic position did not gain enough momentum to deter Germany and other European states (France, Italy) from participating in Nord Stream.

In addition to the WWII references, both opponents and proponents of the pipeline used the Russia–Ukraine gas crises of 2006 and 2009 as a reason why the pipeline should not (or should) be built. The critics described the crises as Russia's one-way manipulation of natural gas price, in order to scare and punish Europe: 'Russia shut down a pipeline that crossed Ukraine, ostensibly over a dispute with Ukraine on pricing and tariff fees. The shutoff left hundreds of thousands of homes in southeastern Europe without heat and shuttered hundreds of factories for three weeks' (Kramer, 2009).

The pro-Russian voices, on the contrary, pointed to Ukraine's multi-billion debt to Russia and claimed that Ukraine had siphoned off natural gas intended for Western Europe. The Russia–Ukraine natural gas trade had been one of the post-Soviet economic distortions: in 2011, when Nord Stream was implemented, Ukraine was paying $280 per 1,000 cubic meters of gas, compared to Germany's price of $330 (Choursina, 2011).

After the Ukraine–Russia disputes, many experts agreed that even though it would be cheaper to upgrade the pipeline that goes through Ukraine, this option became unfeasible after the supply of gas to Europe had been interrupted in the middle of winter. Building another pipeline through the Baltic States and Poland would be equally problematic because of hostile rhetoric between those states and Russia. As a result, Germany and Russia, in September 2005, signed a $6-billion agreement to build Nord Stream (Stent, 2007, p. 426).

The Germany–U.S.–Russia triangle and NATO

Germany, as a major player in European politics, has sometimes been a mediator between Russia and the United States in debated issues, such as the Iranian nuclear program (Szabo, 2009, p. 23). This mediation has often been considered pro-Russian by American analysts, such as the Heritage Foundation and similar organizations ('The Hitler–Stalin Deal Dividing Europe 70 Years Later: How the 1939 Molotov-Ribbentrop Pact Impacts the World Today', 2009). They also argued that Russia was a 'challenger in U.S.–Germany relations' (Szabo, 2009, p. 37).

Germany and the United States also differ in their interpretation of how the Cold War ended and of the energy dependency on Russia. Germany's view has been that the end of the Cold War was a peaceful outcome of negotiations that led to the unification of Germany. The prevailing view in the United States is that the Cold War ended by an overwhelming U.S. *victory* over the Soviet Union (Szabo, 2009, p. 25). Karsten Voigt, a Germany–U.S. coordinator at the German Foreign Ministry, compared the debate around the Nord Stream project to Germany's oil pipeline from the Soviet Union via Poland, built in the 1980s, despite objections by the United States after the Soviet invasion in Afghanistan ('Noviy Evropeisky Miroporyadok: Initsiativi i Strategii Ikh Realizatsii. Sokraschennaya Stenogramma', 2009, p. 86). Voigt emphasized that Germany views the relations with Russia as a *mutual* dependency (including Russia's dependency on German technology and trade). Voigt compared Russia's reaction to the expansion of NATO to Germany's reaction to the European neighbors' policies after WWI, intended to weaken Germany ('Noviy Evropeisky Miroporyadok: Initsiativi i Strategii Ikh Realizatsii. Sokraschennaya Stenogramma', 2009, p. 88).

Energy security has become a theme in discussions about the role of NATO, which is now more than a military security alliance. Some experts have suggested that energy security should also be included in NATO's priorities. For example, Andrew Monaghan of the NATO Defense College proposed that NATO should protect energy businesses and pipelines from military and political threats in situations such as the dispute between Ukraine and Gazprom in January of 2006. After this dispute, energy security appeared for the first time on the agenda of a NATO summit that took place in Latvia in November of 2006 (Monaghan, 2009, pp. 64–65). Monaghan admitted, however, that energy security was a challenging issue to put on NATO's agenda because NATO members themselves differed in their opinion on energy supply options (Monaghan, 2009, p. 71).

Russia has long been extremely dissatisfied with NATO's expansion to the East but still tried to become a part of the EU security framework. Visiting Germany in 2009, Dmitry Medvedev suggested a new architecture of European security that would take into account Russia's role as a major power in Eurasia, pointing to the diminishing power of the United States as a global hegemon and the low functionality of the existing European security system (Kortunov, 2009, p. 36). However, new NATO members have openly expressed fear of Russia (which made them join NATO in the first place), especially after the Georgia–Russia military conflict in August of 2008.

One of the salient aspects of that conflict was a possibility of Georgia and Ukraine becoming NATO members. Angela Merkel was the strongest opponent in the EU of NATO membership for Ukraine and Georgia. Pointing to the ongoing territorial dispute with South Ossetia and Abkhazia, she said that 'countries that are themselves entangled in regional conflicts, can in my opinion not become members' ('Merkel Against NATO Membership for Georgia, Ukraine', 2008).

Energy security has been a divisive issue in tri-lateral Germany–Russia–U.S. relations. At a conference in 2009, Germany stressed that energy supply was the main area in which Russia and the West needed to achieve common understanding. As a solution, German Foreign Minister Steinmeier suggested to create a common security system from Vancouver to Vladivostok, coordinating among the Organization for Security and Co-operation in Europe, the United Nations, the North Atlantic Treaty Organization, and the European Union. At the same conference, Anatoly Korobeinikov of the Russian delegation accused the European Union as being too dependent on the United States, arguing that the expansion of the American influence in Europe prevented the EU from achieving a fuller integration of interests with Russia ('Noviy Evropeisky Miroporyadok: Initsiativi i Strategii Ikh Realizatsii. Sokraschennaya Stenogramma', 2009, pp. 77–82). Back then, Russia presented its energy ties with Europe as a 'soft' economic leverage. In the United States, however, Russia's tri-lateral initiatives with France and Germany were perceived as an attempt to play the 'European' card against the United States, especially during disagreements between the United States and Europe (Stent, 2007, p. 420).

Disagreements between the United States and Germany on the issue of the Nord Stream pipeline were similar to the situation in the 1970s, when the Reagan administration objected to pipelines from Russia's Urengoy gas field to Europe and even banned American companies from selling equipment for these pipelines. Back then, the United States announced that the pipelines were a security threat and increased the over-dependence on Russian energy resources. The Western European states, however, did not have the same concerns and considered the issue to be economic rather than political (Orban, 2008, p. 1).

During the U.S.–Russia 'reset' of 2009, the U.S. government toned down the criticism against the Nord Stream pipeline. The rhetoric about Russia's 'energy weapon', prevalent during the Bush administration, was replaced by discussions about possible cooperation. The U.S. special envoy for Eurasian energy, Richard Morningstar, summed up the sentiment at the time:

> We don't want to have a highly politicized, 'us vs. them' discussion with the Russians…We want to engage with Russia constructively. They are and will continue to be an important player in world energy markets.
>
> (Chazan, 2009)

The U.S. government also stopped public criticism of another pipeline, South Stream, that was on the table at the time of the 'reset' (Chazan, 2009).

To sum up, despite all the potential geopolitical difficulties between Germany and Russia, the two countries have still partnered in energy ties, increasing import–export operations and building the Nord Stream pipeline. The main reasons for such close economic relations have been personal friendships and affinity between the countries' leaders, a long history of keeping economic partnership even in the Cold War period, and separating political and ideological disagreements from economic benefits of import–export, in which Germany provided advanced machinery to the USSR, and the Soviet Union could export raw materials to Germany. It also helped that the USSR and East Germany went through similar transformations after the breakup of the Soviet Union. As a result, Germany's position was often divergent from the other NATO members on such issues as a NATO membership for Georgia and Ukraine, as well as seeing importing natural gas from Russia.

The geopolitical component in this case has been determined by the governments' preferences and strong promotion of their interests via business deals. The level of geopolitical rivalry between Russia and Germany has been low, underlined by substantial initial concessions that Mikhail Gorbachev made during Germany's reunification. As a result of these geopolitical factors, the level of cooperation between Russia and Germany has been very high, especially in the energy sector, but also in other sectors. However, in addition to geopolitical challenges that were solved between Russia and Germany, some domestic interest groups (e.g. environmental NGOs and others) were strongly opposed to the Nord Stream pipeline.

Domestic interest groups

In the case of Russia–Germany energy ties, similar to the previously discussed cases of U.S.–Russia and U.S.–Azerbaijan, governments strongly influenced decision-making in the energy sector. As a result, the domestic groups that align their goals with the government policies have often been more successful in achieving their goals. The controversial Nord Stream pipeline, as well as other aspects of Russia–Germany oil and gas cooperation, has been affected by three major societal groups: German political parties and the media aligned with them; the business lobbies; and environmental and ethnic NGOs.

German political factions: views on Russia

Inside the German political establishment, there has been a substantial debate about how to deal with Russia. Even though views on Russia differ among Germany's parties, they have reached what they call a pragmatic decision of securing as many energy suppliers as possible, including the Nord Stream pipeline ('Mi Izdadim Germano-Rossiiski Uchebnik Istorii: Interview with Walter Jürgen Schmidt', 2009). This consensus resulted in a unified view on Nord Stream. The analysis below looks at the German parties and their leaders at the time when the Nord Stream pipeline was being decided and advocated.

Germany's political scene includes conservative parties (Christian Democrats-CDU and Free Democrats), left-leaning parties (Social Democrats), and environmentalists (Green Party) ('Europe Pushes for Russia to Resume Gas Deliveries', 2009). Among them, Schröder's Social Democratic Party (SDP), led in 2009 by Foreign Minister Steinmeier and Kurt Schumacher, not surprisingly, was very pro-Russian and pushed for stronger cooperation (Szabo, 2009, p. 32). However, former German Foreign Minister Joschka Fischer said later in 2011 that 'it is very unfortunate that we receive a large portion of our energy supply from countries with questionable human rights records' (Steininger, 2011).

Among German parties, the Christian Democratic Union has been very powerful over the last several decades, currently having a CDU incumbent in power and shaping relations with Russia. The CDU generally believes that there are strong economic reasons to cooperate with Russia. The most influential voice in this party – Angela Merkel – takes a cautiously optimistic view, emphasizing the need for multi-dimensional cooperation not only in the energy sector but also in car manufacturing, trade, and human rights promotion (Szabo, 2009, pp. 32–33).

The views on the USSR and then Russia by this powerful political party have been consistent over several decades, even when they diverged from the United States and other NATO partners. In the 1970s, Germany increased economic ties with the USSR despite American opposition. For instance, the 2,000-km Mannesmann natural pipeline with a maximum capacity of 52 billion cubic meters was built from the USSR to West Germany. Another example was the 2,000-km Yamal pipeline in the 1980s, strongly opposed by the American President Ronald Reagan: nevertheless, Chancellor Helmut Kohl proceeded to cooperate with the Soviet Union and made a decision to build the pipeline. The German government put energy security priorities higher than the commitments to isolate the Soviet Bloc.

Around the time when the Nord Stream pipeline was built, the CDU was changing its stance on Russia. In the second half of the 2000s, even though there was some divergence between a cautious and increasingly skeptical incumbent Angela Merkel and the Minister of Foreign Affairs Frank-Walter Steinmeier, 'Steinmeier was the driving force behind Gerhard Schröder's integrative Russia policy' (Meister, 2012). Steinmeier and his ministers promoted the policies of 'rapprochement through interweavement' and 'partnership for modernization' toward Russia, with a great emphasis on energy security and climate change (Meister, 2012). The Christian Democratic–Liberal government coalition of 2009 brought 'sobriety' into Russia–Germany relations, as opposed to personal friendship that had existed between Boris Yeltsin and Helmut Kohl, and between Gerhard Schröder and Vladimir Putin. In 2008–2011, Angela Merkel preferred to deal with President Medvedev rather than with Prime Minister Vladimir Putin. New Liberal Foreign Minister Guido Westerwelle also had other priorities in Eastern Europe, distancing himself from Russia, as Russia lost a part of its importance in the German foreign policy (Meister, 2012).

In the beginning of the current decade, analysts distinguished two different dimensions of the German political rhetoric on Russia. The first one was dissatisfied with Russia's lack of democratic institutions and its human rights violations, and Angela Merkel was a vocal critic of Russia's lack of institutions. The second school of thought was more traditional for CDU and for the Green Party and focused more on purely economic interests for German businesses and German energy security (Meister, 2012).

Political parties' views are best reflected in newspapers that align themselves with these parties and thus shape German public opinion, including views on cooperation with Russia. Major newspapers in Germany are traditionally categorized as center-left, right, and center-right media. When the Nord Stream was debated, the center-left *Süddeutsche Zeitung* took a pragmatic approach and urged a focus on the economic rationale in energy contracts with Russia and transit countries. The newspaper argued that the 'prudent thing to do is to not allow ourselves to be drawn into taking sides [Russia or transit countries] and thereby further the political goals of either of them' ('The Pipeline Power Play', 2009). Therefore, this view emphasized the need to negotiate and follow the business model of cooperation, rather than participate in political debates between Russia and transit states.

The right-leaning *Die Welt* took an almost opposite view and discussed relations with Russia and Russia's energy policies in terms of a geopolitical war. In particular, the newspaper described Russia's natural gas trade as 'a weapon to collect debts but also to sour the West on making Ukraine a member of NATO' ('The Pipeline Power Play', 2009). Referring to Russia's share of natural gas supplies to Germany, the newspaper called Russia's gas trade 'a monopoly'. As a result, the newspaper insisted on Germany's reducing imports of Russian gas and increasing the domestic production of nuclear energy ('The Pipeline Power Play', 2009).

The center-right *Frankfurter Allgemeine Zeitung* acknowledged the importance of the energy market fundamentals, such as the need to pay for oil and gas imports on time and in full and Russia's right to demand a market price from its consumers, including Ukraine. However, the newspaper also argued that price disputes with transit countries also had a political meaning, especially in light of a possible NATO membership for Ukraine ('The Pipeline Power Play', 2009). As a result, the newspaper asserted that 'alternatives must be found' to the Ukrainian pipeline that 'evidently doesn't work' some of the time. Still, according to this view, Nord Stream would increase the EU's dependence on Russian gas. The newspaper brought up the question of whether Iranian gas imports through Nabucco would be more secure than Russian gas supplies ('German Mideast Proposal Doesn't Go Far Enough', 2009).

German political parties and their media outlets had a divergence of view on Russia: some were cautious; others were more optimistic that a mutual dependency would guarantee the stability of oil and gas supplies. The divergence did not result in a wide dissent, however, and the incumbent CDU party proceeded with the line that Germany had been pursuing since the 1969 rapprochement – engaging Russia in economic ties, despite disagreements on political and ideological issues. In this case, German domestic groups that dissented with the government line had not been able to block the Nord Stream pipeline.

Energy business lobbies

Germany and Russia are among the most important trade partners in each other's rankings of trade partners. When the Nord Stream pipeline was finished, in 2011, the two countries increased overall trade volume by 30 per cent, and the annual trade volume reached 75 billion Euro (Meister, 2012).

Germany has various business associations that promote German businesses abroad, including Russia, and German government is aware of their activities and helps them operate successfully. For example, during their trips to Russia, German government officials help German business representatives receive access to the Russian government during intergovernmental meetings (Meister, 2012). It has been especially relevant in Russia, in which vast oil and gas revenues have been conducive to the Russian government's control, merging the Russian political elite with the country's energy sector – 'a rent-distribution mechanism benefiting numerous inside players' (Orttung, 2009, p. 64). Examples include such powerful politicians as Vladislav Surkov, Igor Sechin, Sergei Naryshkin, Aleksey Kudrin, Viktor Khristenko, Yevgeniy Shkolov, Sergei Sobianin, German Gref, and Dmitry Medvedev.

There have been multiple business organizations in Russia, formed to represent the interests of foreign investors. One of them, the Foreign Investment Advisory Council (FIAC), was established in 1994 'to assist Russia in forging and promoting a favorable investment climate based on global expertise and the experience of international companies operating in Russia' (FIAC, 2010). Council members consist of CEOs of foreign companies in Russia, such as ENI S.p.A., Ernst & Young, ExxonMobil, Shell, and Total. Membership is rotated in compliance with the Regulations on the Rotation of Members of the Foreign Investment Advisory Council (FIAC, 2010). This organization is under the Russian government's control, and the Prime Minister of Russia is the chair of the council. FIAC members – 42 foreign companies – comply with the organization's rules and procedures, such as doing charity work in Russia and refraining from decreasing their investments in the country. Membership in this organization is important for companies, as they establish solid connections with the Russian government in order to 'solve problems that foreign investors might have during their investment project implementation via the executive branch and federal government' (Shapovalov, Butrin, & Mazanova, 2010).

German businesses on the Russian market also unite in business associations, focused on European and German representation in Russia ('Rossiisko-Germaniskaya Torgovaya Palata – Predstavitelstvo Nemetskikh Firm v Moskve', 2010). Among them, the German–Russian Forum (Deutsch-Russisches Forum), the European–Russian Business Association (ERBA), and the Association of European Businesses in Russia (AEB) have been most influential. Another organization that specifically focuses on Germany – the Russian–German Trade Chamber (Deutsch-Russische Auslandshandelskammer) – was formed in 2007 on the basis of the German Economic Union in Russia. The Chamber facilitates business contacts between Germany and Russia and provides business services to German companies in Russia.

Russian energy companies unite in their own organizations, such as the Russian Union of Industrialists and Entrepreneurs, the National Energy Security Fund, and the Center of Political Technologies. An organization most relevant to the natural gas industry – the National Association 'Russian Gas Society' – was established in 2001 and now includes 147 natural gas companies. It serves as a liaison for the natural gas industry, promoting the interests of participating Russian companies on the domestic market and globally. The Society represents Russia at major oil and gas forums and conferences around the world, such as the Russia–Europe Energy Dialogue (Starikov, 2010, p. 4). According to its mission statement, the Society communicates with international organizations (e.g. the International Gas Union and the European Gas Industry Union) on behalf of Russian gas producers in legislative and economic issues, defending their interests on international markets (especially during trade disputes) and conducting public relations on behalf of the industry. Traditionally, the position of this organization has been strongly aligned and intertwined with the Russian government's energy policy.

Government appointees lead the Russian gas industry. Over the last decade, Gazprom's more independent managers were replaced by those with close ties to the Kremlin. One example of such a resignation was Alexander Ryazanov, deputy CEO of Gazprom and President of Gazprom Neft, who was considered excessively 'independent' in his actions. He insisted that Gazprom Neft remain a separate company rather than being integrated with Gazprom. He also disagreed with Gazprom's CEO Alexey Miller about the TNK-BP deal and oil export volume ('Gazprom Uvolil Alexandra Ryazanova za Samostoyatelnost', 2006). As a result, Ryazanov was replaced with Valery Golubev, who had worked in the KGB for more than a decade and later at St. Petersburg Mayor's office with Vladimir Putin ('Gazprom Uvolil Alexandra Ryazanova za Samostoyatelnost', 2006).

In 2006, when the Russian government was strengthening its control over the oil and natural gas industry, nine out of fifteen members on Gazprom's Board of Directors were from St. Petersburg: CEO Alexey Miller, Chief Accountant Elena Vasilyeva, Chief of Finance Andrei Kruglov, Chief of Security Sergey Ushakov, Chief of Construction and Investments Valery Golubev, Chief of Asset Management and Corporate Relations Elena Pavlova, Chief of Marketing and Refining Kirill Seleznev, Chief of Law Department Konstantin Chuichenko, and Deputy CEO Alexander Kozlov ('Alexander Ryazanov ne Vkhodil v "Piterskoe" Bolshinstvo Pravleniya "Gazproma"', 2006). Another close friend of Vladimir Putin since the 1980s – Nord Stream AG Managing Director Matias Varnig – was lobbying for Nord Stream in Europe (Ispolnova, 2009).

In sum, Russian energy business lobbies have been very tightly intertwined with the government. They may be considered an extension of the government's foreign policy, rather than a separate force on the international arena. Domestically, they are extremely powerful in obtaining the most favorable government contracts for the oil and gas industry. In the case of Nord Stream, as in cases of other pipelines, the decisive voice belonged to the participating governments at the level of heads of state. Private and state controlled energy businesses had to follow the government lead.

Think tanks and environmental NGOs

Think tanks and environmental NGOs, as social entities less intertwined with the government than the political parties and energy business lobbies explored above, have been the most vocal critics of the Nord Stream pipeline, mobilizing and channeling the grassroots public voice. While think tanks and NGOs in Russia did not ultimately have a substantial influence on the Nord Stream pipeline planning and implementation, the European think tanks and NGOs were much more active players in debates about the pipeline.

German think tanks raised concerns against the German government's decision to build the Nord Stream pipeline. Unlike American think tanks, however, the German ones focused less on the geopolitical dangers of Russia's domination and more on the need for Germany to diversify sources of energy as much as possible. Oliver Geden of the German Institute for International and Security Affairs (a policy-oriented think tank established in 1962) warned that excessive dependency on energy producing countries subjected Germany to the risk of being blackmailed by the producers. As a solution, Geden suggested that Europe should create an internal energy gas market that would include natural gas and other sources of energy: 'The EU can achieve more for its gas supply security by expanding its internal energy market than it could by building any new import pipeline' (Geden, 2012). He further predicted that Germany's climate change policy (which was prioritizing renewable energy sources) might even make natural gas itself a 'thing of the past' in the coming decades. In the meantime, he argued that competition between the two pipelines that were discussed after Nord Stream had been built – the Nabucco and South Stream – was mainly geopolitical, in which each side (Turkey for Nabucco and Russia for South Stream) tried to minimize the economic justification for the rival pipeline. Geden concluded that the key to the European energy security was the maximum diversification of energy suppliers and energy sources, as well as the creation of the unified European internal energy market (Geden, 2012).

Another think tank – the German Institute for Economic Research (DIW Berlin) – expressed a similar view about the need to diversify energy sources as broadly as possible ('Germany, Russia to Launch Nord Stream', 2011). Claudia Kemfert of this institute emphasized the role of new technology and the upcoming global market of LNG, in which suppliers will be competing with one another: 'There is a gas surplus in the world – why fixate on Russia?' ('Germany, Russia to Launch Nord Stream', 2011).

The German Institute for Economic Research concluded that Nord Stream was an important step for Europe in diversifying energy routes. The analysts of this institute examined costs and predicted profits of the pipeline and argued that Nord Stream was actually the most expensive way to deliver natural gas to Germany. They continued that the decision to build the pipeline had been made based on the strategic grounds, mainly to avoid the transit states – Ukraine, Poland, and Belarus (Hubert & Suleymanova, 2009). Overall, the institute still came to the conclusion that the Nord Stream pipeline would have positive implications for European energy security, bringing natural gas as a transition fuel, despite concerns about overdependence on Russia, if Europe continues to diversify its energy mix in the long term

Not unlike the previous two think tanks, the Euro Institute for Information and Technology Transfer in Environmental Protection agreed that multiple pipelines, including Nord Stream, South Stream, Nabucco, and others were 'urgently needed' ('Pipeline Technology Conference 2010 – Final Report', 2010). The Institute also welcomed more LNG supplies to Europe. Similarly, the Institute of Transportation Systems focused on the technical side of safe transportation of energy resources around Germany, including railway systems.

Some think tanks, more focused on political analysis, traditionally point to possible resource wars as oil and gas become depleted. Alexander Rahr of the German Council on Foreign Relations predicts that 'the twenty-first century will be the century of fights over natural resources,... and we will see new alliances struck all over the globe attempting to create energy security' (Steininger, 2011). Around 2011–2012, multiple think tank experts in Germany expressed their opinions and concerns about the future of relations with Russia.

Stefan Meister of DGAP's Robert Bosch Center for Central and Eastern Europe, Russia, and Central Asia for Eastern Europe, Russia and Central Asia provided an insightful geopolitical analysis of Russia–Germany relations and predictions for the future course after Vladimir Putin came back to presidency for the third term in 2012 (Meister, 2012). Meister argued that the goals of Russia and Germany in the course of cooperation on Nord Stream were quite different: 'While the German side wants to develop common projects of good practice which will modernize the Russian economy and politics, the Russian side is interested in technology transfer' (Meister, 2012). At the time, though, Russian President Dmitry Medvedev was also trying to bring modernization to the Russian economy – a technological breakthrough similar to Silicon Valley in the United States. Meister convincingly showed that Russia–Germany relations were changing, as the anti-democratic turn of the Russian political development in 2012 was not going to find support in Germany at all levels.

Hubertus Bardt of the Cologne Institute for Economic Research warned again about gas disputes between Russia and Ukraine: 'There was no bad guy and good guy in the gas conflict between Russia and Ukraine... They both played a high-risk game. But it showed Europe how vulnerable it is' (Steininger, 2011). He took a cautiously optimistic view about Russia, though, saying that Russia proved 'to be a very reliable supplier, even in Cold War times. Without selling gas, Russia's whole economy and its social security would collapse' (Steininger, 2011).

In addition to think tanks, Nord Stream attracted the attention of multiple environmental NGOs that were concerned about the impact of the pipeline on the marine environment of the Baltic Sea. The NGOs were concerned about the pipeline's possible damage to marine wildlife and disturbing toxic wastes and chemical munitions, remaining on the sea bottom since WWII. NGOs investigated issues and problems that needed to be solved during the pipeline's construction and maintenance: corrosion, leaks, accident prevention, and protection from terrorism.

Most importantly, chemical weapons on the bottom of the Baltic Sea, dumped after the end of WWII, could endanger the pipeline and damage the surrounding natural habitat (Hubert & Suleymanova, 2009). No consensus had been reached in the scientific community about what needed to be done about those chemical weapons on the bottom of the sea. Leaks from chemical munitions in the Baltic Sea could affect the population of fish and hurt the fisheries that are a vital component of the surrounding countries' economies. Fishermen could become injured by chemical substances spilled from munitions in the sea or by old weapons caught in fishnets. A closed environment and the level of pollution of the Baltic Sea made pipeline construction and maintenance very challenging, forcing the government to look for a balanced solution between the need for energy sources and the preservation of the environment (Hubert & Suleymanova, 2009).

Not surprisingly, environmental NGOs, both in Europe and Russia, were extremely vocal about the Nord Stream pipeline's possible environmental effects. Greenpeace, Green Cross, WWF, the Baltic Fund for Nature, among others – all of them raised concerns about future damage to fisheries, the disturbances of WWII-chemical munitions, possible leaks and accidents, and the compensation to those people whose lifestyle would be destroyed by the industrial infrastructure that went together with the pipeline. Russian environmental NGOs blossomed in the early 1990s, as a result of democratic reforms and international NGOs coming to open branches in Russia (Perovic et al., 2009). They have been active in their commentary and activism in the 2000s as well, although the strengthening of the state power in Russia and the setback in democratic reforms reduced their influence to an advisory and consultant role, lacking a decision-making voice in multi-billion-dollar projects, such as Nord Stream.

Despite the fact that NGOs were not able to block the decision to build the pipeline, they achieved quite a remarkable goal nevertheless – to hold large energy companies more accountable on the environmental front. In 2008, the Nord Stream managing company conducted a comprehensive environmental assessment of the pipeline and regularly held meetings with NGOs before the construction was started, documenting all meetings on the company's website. When Nord Stream decided to modify the route of the pipeline on the Gogland Island, the Forum of Baltic NGOs, the international coalition 'Clean Baltics', the Russian Regional Environmental Centre (RREC), the Baltic Environmental Forum Group (Riga), Project 'Ecocentrum' (Saint-Petersburg), and other organizations were given explanations concerning the threats to the Ingermanlandsky nature reserve located in that area. Alexey Yablokov (Boris Yeltsin's Counselor for Ecology and Chair of the Russian National Security Council's Interagency Commission for Ecological Security in the early 1990s, Minister of Environment and then Minister of Economy in 1992–1993, and later President of the Centre for Russian Environmental Policy, and correspondent-member of the Russian Academy of Sciences) was one of the experts who raised strong concerns about Nord Stream's environmental consequences, especially related to chemical munitions. He pointed to the lack of transparency by the Nord Stream management in addressing these concerns (Yablokov, 2008).

The case of Nord Stream made large companies in Russia more accountable to the NGOs and the public in environmental issues. Nevertheless, NGOs were not able to block projects implemented by the Russian government, because the Nord Stream project had high geopolitical and economic importance and was pushed by the governments.

Conclusion

Despite Germany's leading position in the European Union, Germany has not always followed the European policy of reducing Russia's share of natural gas supplies. On the contrary, Germany and Russia recently constructed a multi-billion-dollar natural gas pipeline on the bottom of the Baltic Sea – Nord Stream – that directly connects the two countries, bypassing traditional transit states (Poland, the Baltic States, Belarus, Ukraine). This pipeline project is puzzling because it possibly increased Germany's (and European) dependency on Russian natural gas, which already was extremely high. The pipeline has been criticized by many European governments who lost leverage and transit fees, by environmental NGOs, by German think tanks, and even NATO allies, such as the United States. Nevertheless, Germany proceeded with the project that received the highest approval on the government level.

The geographic advantage has been vital in this case – Russia is the biggest producer of natural gas in the world, and Russian gas is still cheaper for Germany than LNG from the Middle East or the United States. Germany has been aware of its vulnerable position in energy security throughout the decades and has practised a pragmatic approach to the USSR and then Russia even in the period of the Cold War, and especially after the breakup of the Soviet Union.

Although the economic rationale is the strongest element in this case study, both the Russian and German governments fully controlled the planning and implementation of the Nord Stream project, which is in line with the geopolitical explanation. In this case, geopolitical considerations were also extremely influential, as Germany achieved an unprecedented access to Russia's energy resources and increased its influence on the EU gas redistribution and storage market. Russia, in its turn, satisfied its long-term loyal customer for Russian natural gas.

References

Aalto, P. (2008a). The EU–Russia Energy Dialogue and the Future of European Integration: From Economic to Politico-Normative Narratives. In P. Aalto (Ed.), *The EU-Russian Energy Dialogue: Europe's Future Energy Security* (pp. 23–41). Hampshire: Ashgate Publishing.

Aalto, P. (Ed.). (2008b). *The EU–Russian Energy Dialogue: Europe's Future Energy Security*. Hampshire: Ashgate Publishing.

Alexander Ryazanov ne Vkhodil v 'Piterskoe' Bolshinstvo Pravleniya 'Gazproma'. (2006, November 17). *Au92*. Retrieved from http://www.au92.ru/msg/20061117_bonmbcf.html

Anon. (2011). The Start of New German–Russian Cooperation. *Stratfor Global Intelligence*. Retrieved from https://www.stratfor.com/analysis/start-new-german-russian-cooperation

Assenova, M. (2011). Bulgaria Terminates a Russian-led Energy Project, Discovers Natural Gas and Prospective Shale Gas Deposits. *The Jamestown Foundation*. Retrieved from http://www.jamestown.org/programs/edm/single/?tx_ttnews[tt_news]=38791&tx_ttnews[backPid]=27&cHash=58cb79d651634b72687e7419e0650e05

Bandlova, T. (2010). Rossiya-Germania: Nadyozhnoe i Bespereboinoe Obespechenie Siryom. [Russia–Germany: Reliable and Uninterrupted Supply of Commodities]. *Gazoviy Business*, 2, 10–12.

Bierman, S., & Meyer, H. (2009, January 3). Russia, Ukraine dig in heels in gas dispute. *The Globe and Mail (Index-only)*.

Buerrieri, F. (2012). EU–Russia Relations: A Chance Not To Be Missed. *European Alternatives*. Retrieved from: http://www.euroalter.com/2010/eu-russia-relations-a-chance-not-to-be-missed/

Bulgaria Oil & Gas Report – Q2 2015. (2015) (pp. 1–117). London: Business Monitor International.

Chazan, G. (2009, November 27). World News: U.S. Gears Back Criticism of Two Russia-Backed Pipelines. *Wall Street Journal*.

Choursina, K. (2011, February 16). Ukraine to Pay More for Russian Natural Gas, Interfax Reports. Retrieved March 5, 2011, from http://www.bloomberg.com/news/2011-02-16/ukraine-to-pay-more-for-russian-natural-gas-interfax-reports.html

Costs for Germany's Nuclear Exit Might Increase to 70 Billion Euros. (2015). *Transmission & Distribution World*.

Daly, J. C. K. (2011). The Impact of Germany's Decision to Phase Out Nuclear Energy. *Journal of Foreign Relations*, (November 11).

Energy Infrastructure: What Do We Want to Achieve? (2008, November 12). Retrieved March 25, 2009, from http://ec.europa.eu/energy/infrastructure/index_en.htm

Europe Fails to Wean Itself Off Russian Gas. (2009). Retrieved January 12, 2009.

Europe Pushes for Russia to Resume Gas Deliveries. (2009, June 5). *Spiegel Online*.

European Market. (2011). Distrigas.

FIAC. (2010). About FIAC. Retrieved from: http://www.fiac.ru/about.php

Filatova, I. (2010). Siemens Signs Billions of Dollars of Russian Deals. *The Moscow Times*, (4442). Retrieved from http://www.themoscowtimes.com/business/article/siemens-signs-billions-of-dollars-of-russian-deals/410475.html

Gas Exporting Countries Forum Seeks Cooperation with Azerbaijan. (2014, November 29). *BBC Monitoring Central Asia*.

Gas; France Clears Total to Work at Yamal LNG, Kharyaga and Termokarstovoye. (2015, April 28). *Interfax: Russia & CIS Energy Daily*.

Gazprom. (2011). Gazprom in Questions and Answers. Retrieved March 2, 2011, from http://eng.gazpromquestions.ru/?id=4

Gazprom Losing Battle for World LNG Market. (2010). *Oil and Gas Eurasia*, (August 5, 2010). Retrieved from: http://www.oilandgaseurasia.com/news/p/0/news/8250

Gazprom Opredelilsya s Tsenoi na Gas Dlya Evropy: Ona Budet Nizhe, Chem Dlya Urkainy. (2012). *Korrespondent.Net*. Retrieved from http://korrespondent.net/business/companies/1317485-gazprom-opredelilsya-s-cenoj-na-gaz-dlya-evropy-ona-budet-nizhe-chem-dlya-ukrainy

Gazprom Says European Gas Deliveries Could Resume Friday. (2009, January 9). *Spiegel Online*.

Gazprom Uvolil Alexandra Ryazanova za Samostoyatelnost. (2006, November 16). *Lenta. Ru*. Retrieved from: http://www.lenta.ru/news/2006/11/16/ryazanov/

Geden, O. (2012). EU and Russia: The Pipeline Race Delusion. *EUObserver.Com*, (August 2). Retrieved from: http://euobserver.com/foreign/115165

Gelb, B. A. (2007). CRS Report for Congress – Russian Natural Gas: Regional Dependence. Congressional Research Service, The Library of Congress.

German Commentary Finds Putin, Merkel 'Cool' Towards Each Other. (2006, October 11). *BBC Monitoring European*, p. 1.

German Mideast Proposal Doesn't Go Far Enough. (2009, January 12). *Spiegel Online*.

Germany, Russia to Launch Nord Stream. (2011, November 6). *The Local: Germany's News in English*. Retrieved from http://www.thelocal.de/national/20111106-38697.html

Green, S. (2010). EU-Russia: You Can't Always Get What You Want. (June 17, 2010). Retrieved from http://www.carnegieendowment.org/publications/index.cfm?fa=view&id=41009

Grib, N. (2009, December 14). Gazprom Smog Obognat Tolko Nigeriyu po Dinamike Postavok Gaza v Evropu. *Kommersant*, p. 11.

Grigoriev, L., & Milov, V. (2009). *Energetika Kak Instrument Politiki*. Moscow: Radio Echo Moskvi.

Gudkov, A. (2009, December 14). Belorussia Podpisala Neftyanoi Schet: Eksportnie Poshlini RF Konvertiruyutsya v Aktsii Belorusskih Kompanii. *Kommersant*.

Gurkov, A. (2010). Leonid Grigoriev: Spros na Energiyu iz Rossii nachnyot Sokraschatsya v Evrope let cherez 30–40. *InoSMI.ru*. (July 21, 2010). Retrieved from http://www.inosmi.ru/russia/20100721/161490654.html

Guseinov, V. (2008). *Politica u Poroga Tvoego Doma. Izbrannie Publitsisticheskie i Analiticheskie Statji*. Moscow: Krasnaya Zvezda.

Hanson, P. (2009). The Sustainability of Russia's Energy Power: Implications for the Russian Economy. In J. Perovic, R. Orttung, & A. Wenger (Eds.), *Russian Energy Power and Foreign Relations* (pp. 23–50). London: Routledge.

Hellfeld, M. v., & Chase, J. (2010, July 14). How Kohl and Gorbachev Sealed the Deal On German Reunification. *DW*. Retrieved from http://www.dw.de/dw/article/0,,5788998,00.html

The Hitler-Stalin Deal Dividing Europe 70 Years Later: How the 1939 Molotov-Ribbentrop Pact Impacts the World Today. (2009). Washington, DC: The Heritage Foundation.

Hubert, F., & Suleymanova, I. (2009). Weekly Report. Baltic Sea Pipeline: The Profits Will Be Distributed Differently. Berlin: German Institute for Economic Research.

Intellpuke (2009). European Union Pushes For Russia To Resume Gas Deliveries. Free Internet Press. Retrieved on September 8, 2015 from http://freeinternetpress.com/story/European-Union-Pushes-For-Russia-To-Resume-Gas-Deliveries-19727.html

Ispolnova, D. (2009, February 12). Ot VTB s Lyubovyu. *Gazeta.Ru*. Retrieved from http://www.gazeta.ru/financial/2009/02/12/2941393.shtml

Italy/Russia: Italy Faces Dire Situation if Russian Gas Supplies Cut. (2014, August 29). *Asia News Monitor*.

Johnson, D., & Robinson, P. (2005). *Perspectives on EU-Russia Relations*. New York, NY: Routledge.

Kanter, J. (2015, May 11). U.S. Urges Greece to Spurn Russian Gas Pipeline Proposal. *International New York Times*.

Khoroshun, N. D. (2014). Geo-Economic Priorities of Russian Federation and Federal Republic of Germany in the Context of Overcoming the Crisis. *Asian Social Science, 10*(20), 184–194.

Klitsounova, E. (2005). The Russian Perspective. In D. Johnson & P. Robinson (Eds.), *Perspectives on EU-Russia Relations* (pp. 35–54). New York, NY: Routledge.

Koranyi, D., & Vatansever, A. (2013). *Lowering the Price of Russian Gas: A Challenge for European Energy Security*: Atlantic Council of the United States.

Kortunov, S. (2009). Novaya Arkhitektura Evrobezopasnosti. [A New Architecture of European Security]. *Vestnik Analitiki, 4*(38), 34–45.

Kramer, A. E. (2009, October 12). Russia Gas Pipeline Heightens East Europe's Fears. *The New York Times*. Retrieved from http://www.nytimes.com/2009/10/13/world/europe/13pipes.html

Kruk, M. (2012, June 19). Polish Shale Gas Industry to Plod On Even as Exxon Pulls Plug. *The Wall Street Journal*. Retrieved from http://blogs.wsj.com/emergingeurope/2012/06/19/polish-shale-gas-industry-to-plod-on-even-as-exxon-pulls-plug/

Lyutskanov, E., Alieva, L., & Serafimova, M. (2013). *NATO Science for Peace and Security Series – E : Human and Societal Dynamics, Volume 110 : Energy Security in the Wider Black Sea Area – National and Allied Approaches*. IOS Press.

Mazneva, E. (2009, December 14). Gaz v Obmen na Obeschaniya. *Vedomosti*, p. A03.

McCann, K. (2005). EU Technical Assistance, Programmes and Projects: An Assessment of Energy Sector Programmes in Eastern Europe and the Former Soviet Union. In D. Johnson & P. Robinson (Eds.), *Perspectives on EU-Russia Relations* (pp. 194–213). New York, NY: Routledge.

Medvedev Speaks at Nord Stream Gas Pipeline Launch Ceremony – Kremlin Transcript. (2011, November 9). *BBC Monitoring Former Soviet Union*.

Meister, S. (2012, July 7). An Alienated Partnership. *Vestnik Kavkaza*. Retrieved from http://vestnikkavkaza.net/analysis/politics/28706.html

Merkel Against NATO Membership for Georgia, Ukraine. (2008, March 10). *Terra Daily*. Retrieved from http://www.terradaily.com/reports/Merkel_against_NATO_membership_for_Georgia_Ukraine_999.html

Mi Izdadim Germano-Rossiiski Uchebnik Istorii: Interview with Walter Jürgen Schmidt. (2009, August 21). *Gazeta.Ru*. Retrieved from http://www.gazeta.ru/interview/nm/s3234837.shtml

Miller, R. (1992). Are Two Germanies Better Than One? Russo-German Relations: Past, Present, and Future. Retrieved from http://history.eserver.org/russo-german-relations.txt

Monaghan, A. (2009). Energeticheskaya Bezopasnost: Ogranichennaya Dopolnyayuschaya Rol NATO. [Energy Security: A Limited Supplemental Role of NATO]. *Vestnik Analitiki, 1*(35), 64–71.

Morozov, V. (2008). Energy Dialogue and the Future of Russia: Politics and Economics in the Struggle for Europe. In P. Aalto (Ed.), *The EU-Russian Energy Dialogue: Europe's Future Energy Security* (pp. 43–61). Hampshire: Ashgate Publishing.

Muiznieks, N. (2009). Energy Security, Memory Wars and 'Compatriots' in Baltic–Russian Relations. Stanford University.

Newnham, R. (2002). *Deutsche Mark Diplomacy: Positive Economic Sanctions in German-Russian Relations*. Pennsylvania, PA: Pennsylvania State University Press.

Nord Stream AG: Project Information Document. Offshore pipeline through the Baltic Sea (2006) (pp. 1–94). Rambøll A/S and Nord Stream AG.

Noviy Evropeisky Miroporyadok: Initsiativi i Strategii Ikh Realizatsii. Sokraschennaya Stenogramma. (2009). [New European Order: Initiatives and the Strategies of Their Implementation. Conference Minutes]. *Vestnik Analitiki*, *3*(37), 75–104.

Orban, A. (2008). *Power, Energy, and the New Russian Imperialism*. London: Praeger Security International.

Orttung, R. (2009). Energy and State-Society Relations. In J. Perovic, R. Orttung, & A. Wenger (Eds.), *Russian Energy Power and Foreign Relations* (pp. 51–70). London: Routledge.

Papava, V., Grigoriev, L., Paczynski, W., Tokmazishvili, M., Salikhov, M. Sr., & Bagirov, S. (2009). Energy Trade and Cooperation between the EU and CIS Countries. *SSRN eLibrary*.

Perovic, J., Orttung, R., & Wenger, A. (Eds.). (2009). *Introduction: Russian Energy Power, Domestic and International Dimensions*. London: Routledge.

The Pipeline Power Play. (2009, January 7). *Spiegel Online*.

Pipeline Technology Conference 2010 – Final Report. (2010). Retrieved from http://www.eitep.de/de/index.php?option=com_content&task=view&lang=en&id=128

Poland Oil & Gas Report – Q3 2015. (2015) (pp. 1–160). London: Business Monitor International.

Putin Thanks Schröder, All Nord Stream Participants. (2011, September 6). *Daily News Bulletin*.

Razavi, H. (2007). *Financing Energy Projects in Developing Countries*. Tulsa, OK: Penn Well Corporation.

Reluctant Minister Introduces Fracking Bill. (2015, May 7). *Deutsche Welle*.

Robertson, H. (2013). Russia Won't Develop Shale Gas for a Decade. *Petroleum Economist*.

Rossiisko-Germaniskaya Torgovaya Palata – Predstavitelstvo Nemetskikh Firm v Moskve. (2010). Retrieved from http://russland.ahk.de/ru/menu2/ueberuns/

Rossiya Ne Dolzhna Upustit' Svoi Shans. [Russia Should Not Miss Its Chance]. (2009). *Vestnik Aktualnikh Prognozov. Rossiya: Tretye Tisyacheletije*, *22*, 40–41.

Russia Scraps South Stream Natural Gas Pipeline Project. (2014, December 2). *Progressive Digital Media Oil & Gas, Mining, Power, CleanTech and Renewable Energy News*.

Russia to Host Gas Exporting Countries Forum Summit July 1–2. (2013, May 8). *Interfax: Russia & CIS Energy Newswire*.

Russia–Ukraine Agreements; Russia Agrees to Lower Gas Price for Ukraine by a Third, Customs Union Not Discussed at Presidents' Meeting. (2013). *Interfax: Russia & CIS Business & Investment Weekly*.

Russian president calls for improving investment climate. (2010, February 2). *BBC Monitoring Former Soviet Union*.

RWE, Russians Explore Energy Efficiency. (2011, May 16). *UPI.com*. Retrieved from http://www.upi.com/Business_News/Energy-Resources/2011/05/16/RWE-Russians-explore-energy-efficiency/UPI-45031305551343/

Shapovalov, A., Butrin, D., & Mazanova, Z. (2010, February 18). Inostrannikh Lobbistov Vstroili v Vertical. *Commersant*. Retrieved from http://www.lobbying.ru/print.php?article_id=5248

Stack, M. K., & Rotella, S. (2009, January 7). The World; Russia Again Cuts Gas Flow to Europe; Gazprom Accuses Ukraine of Siphoning Fuel from the Cross-country Pipeline Amid the Pricing Dispute. *Los Angeles Times*.

Starikov, A. (2010). Rossiiskoe Gazovoe Obschestvo: Na Poroge Yubileya. [Russian Gas Society: On the Threshold of Anniversary]. *Gazoviy Business*, *2*, 4–9.

Statistics; Belarus Ups Oil Imports 5.1%, Petroleum Product Exports 35.5% in Jan. (2015, April 6). *Interfax: Kazakhstan Oil & Gas Weekly*.

Steininger, M. (2011). Nord Stream Pipeline Opens, Russia-Europe Interdependence Grows. *The Christian Science Monitor*. Retrieved from http://www.csmonitor.com/World/Europe/2011/1108/Nord-Stream-pipeline-opens-Russia-Europe-interdependence-grows

Stent, A. (2000). *Russia and Germany Reborn: Unification, the Soviet Collapse, and the New Europe*. Princeton, NJ: Princeton University Press.

Stent, A. (2007). Reluctant Europeans: Three Centuries of Russian Ambivalence Toward the West. In R. Legvold (Ed.), *Russian Foreign Policy in the Twenty-First Century and the Shadow of the Past* (pp. 393–441). New York, NY: Columbia University Press.

Stern, J. (2005). *The Future of Russian Gas and Gazprom*. Oxford: Oxford University Press.

Stern, J. (2011). Natural Gas in Europe – the Importance of Russia. Retrieved from www.centrex.at/en/files/study_stern_e.pdf

Szabo, S. F. (2009). Can Berlin and Washington Agree on Russia? *The Washington Quarterly*, 32(4), 23–41.

Think Tank: Institute of Energy Policy (IEP). (2010, January 29). Retrieved September 20, 2010, from http://www.russiaprofile.org/resources/ngos/environment/institute_of_energy_policy

Tolstikh, P. A. (2006). *Praktika Lobbisma v Gosudarstvennoi Dume Federalnogo Sobrania Rossiiskoi Federatsii (Lobbyism in the Russian Duma)*. Moscow: Kanon.

Topalov, A. (2009, October 7). 'Severniy Potok' Upyorsya v Dokhod. *Gazeta.Ru*. Retrieved from http://gazeta.ru/business/2009/10/07/3270585.shtml

Turkmenistan believes in Trans-Caspian pipeline project. (2015, May 9). *McClatchy – Tribune Business News*.

Vatansever, A. (2010, June 17). EU-Russia Energy Relations: A Pause or Fast Forward? *Commentary*. Retrieved from http://www.carnegieendowment.org/publications/index.cfm?fa=view&id=41006

Wall Street Sector Selector: Liquid Natural Gas Europe's Answer to Dependency on Russia. (2014). Chatham: Newstex.

Western Europe Oil and Gas Insight – August 2014. (2014) (pp. 1–12). London: Business Monitor International.

Wiegrefe, K. (2010, August 7). How East Germany Tried to Undermine Willy Brandt. *Der Spiegel*. Retrieved from http://www.spiegel.de/international/germany/ostpolitik-how-east-germany-tried-to-undermine-willy-brandt-a-705118.html

World; World Bank: Russia Slips in Doing Business Rank Despite Reforms. (2010, November 8). *Interfax: Ukraine Business Daily*.

Yablokov, A. (2008, October 7). Zachem Pahat' More? *Izbrannoe*. Retrieved from http://www.izbrannoe.ru/41514.html

Yazev, V. (2010). *Russia and the International Energy Cooperation in the XXI Century*. Moscow: Granica.

Zygar, M., & Grib, N. (2009). Mister Neft. *Kommersant*, *78/P*(4133). Retrieved from http://kommersant.ru/doc.aspx?DocsID=1164267&NodesID=4

5 The Ukraine conflict

A game-changer?

Ukraine is turning to Europe

The recent political crisis in Ukraine started in November 2013, when President Viktor Yanukovych decided, under pressure, not to proceed with an EU association agreement. Such pressure included, according to the Western analysts, Moscow's 'banning Ukrainian exports', especially from the Eastern Ukraine, economically dependent on Russia for the region's export revenues ('Senate Foreign Relations Committee Hearing', 2015). Yanukovych himself called this pressure 'blackmail'. In his conversation with Lithuanian President Dalia Grybauskaite, he explained his decision to cancel the EU agreement by the 'necessity to improve considerably trade and economic relations with Russia and the CIS countries'. He announced his decision at the EU summit in Lithuania on November 28–29, 2013 ('Economic Policy; Yanukovych Tells Grybauskaite Kyiv Cannot Sign Agreement With EU Because of Russian Pressure', 2013).

The agreement was supposed to relax trade restrictions and border protection between the European Union and Ukraine. The deal did not go as far as to offer EU membership to Ukraine, but it was the first step for Ukraine to integrate its economy in the European economic system, partially moving the country away from Russia's political and economic influence. The agreement was designed to open both Ukrainian and European markets to easier goods and services movements and potentially to lower visa restrictions for travelers between Ukraine and European countries.

Just a couple of months before the fateful announcement about the halting of the agreement, at the 68th session of the UN General Assembly in New York during September 2013, Viktor Yanukovych had been enthusiastic about 'the realization of Ukraine's European aspirations' that would be also, in his view back then, beneficial for Ukraine–Russia relations and Russia-led Custom Union:

> Our relations with the EU will benefit the Customs Union. Ukraine is a bridge between Russia and the EU and it's very important to make sure the

> bridge is strong and reliable. A dialog between Ukraine, Russia and the EU on trade issues is possible in the near future.
>
> ('Economic Policy; Yanukovych: Dialog Between Ukraine, Russia, EU on Trade Issues Possible in Near Future', 2013).

At that time, Viktor Yanukovych did not, in his own words, foresee any impediments to the upcoming agreement ('Economic Policy; Yanukovych: Dialog Between Ukraine, Russia, EU on Trade Issues Possible in Near Future', 2013). As he admitted later, Russia opposed the agreement and pressured him into not signing the deal with the European Union. In 2013, however, he could not predict the chain of tragic events that would unfold following the abandoning of the EU deal ('Ukraine Protests After Yanukovych EU Deal Rejection', 2013).

Viktor Yanukovych's abandoning the EU agreement divided Ukraine along the lines of future relations with Russia. Right after he announced that the agreement was not going to be signed, about 10,000 disappointed pro-Europe supporters took to the streets of Ukraine's capital, Kiev, to protest the president's decision. Since the pro-European protests first began on the Maidan Square, a central location in Kiev, the media around the globe started to refer to them as Euromaidan protests. Simultaneously, about 4,000 pro-Russia demonstrators gathered to express their support for Viktor Yanukovych's decision, welcoming the continuation of closer economic ties with Russia ('Ukraine Protests After Yanukovych EU Deal Rejection', 2013). The demonstrations started as peaceful but quickly turned into an armed conflict between pro-Russian Eastern Ukraine and the rest of the country with a more pro-European stance. By mid-2015, the conflict led to thousands of casualties in just a year and a half. To quell the pro-European demonstrations that started in November 2013 and lasted for months, on January 15, 2014, President Yanukovych adopted the anti-opposition legislation.

It was not the first time he targeted the opposition movement. Back in 2011, the Ukrainian government charged former Prime Minister Yuliya Tymoshenko, who was at the time the leader of the opposition forces, with fraud related to her misuse of government money while she had been in power. Tymoshenko denied the accusations and claimed that the charges were politically motivated, in order to weaken her opposition leadership role. Other former government officials and civil servants were charged with 'wrongful appropriation by the state' of 11 billion cubic meters of natural gas from the RosUkrEnergo gas company, owned by a close ally of Victor Yanukovych ('Ukraine Politics: Authorities Target Opposition', 2011).

Even though all those charges did not seem, at first sight, politically motivated, the opposition movement voiced concerns that the investigations were used by the Yanukovych administration in order to suppress the dissent in the country. In response, Viktor Yanukovych described the investigations as his anti-corruption policy implementation that he had promised during his presidential campaign. To fulfill his election promises, in December 2011, the government passed an anti-corruption law that was later approved by the Group of States Against

Corruption (GRECO) of the Council of Europe and that replaced the previously existing anti-corruption law of 2009. The 2011 anti-corruption law was complemented by the January 2012 law on public information access, which was an attempt to reconcile with the opposition, since the law had been designed by Yuliya Tymoshenko's supporters ('Ukraine Politics: Authorities Target Opposition', 2011).

Despite all that previous dealing with the opposition, during the pro-European demonstrations of early 2014, President Yanukovych seemed to have lost the control of the situation in Kiev. The Russian government was trying to help him, and in order to appease the pro-Europe protesters in Kiev, Russia promised Ukraine to continue the delivery of cheap natural gas and a loan of $15 billion on favorable terms, hoping that such offers would make the demonstrations calm down. However, the demonstrations continued, and Viktor Yanukovych was unable to find a way to stop the unrest peacefully ('Senate Foreign Relations Committee Hearing', 2015).

The pro-European protesters, inspired by the quickly rising Right Sector movement, refused to leave the Maidan Square. New opposition leaders (Oleh Tyahnybok, Arseniy Yatseniuk, and Vitali Klitschko) were emerging in the movement, but the crowds were organized mostly by the Right Sector. On February 20, 2014, the protesters, armed with Molotov cocktails and stones, started to march on the Ukrainian Parliament, and the riot police were ordered to protect the Parliament (Bilozerska, 2015). The violence started with a single sniper fire at the previously peaceful demonstration. It is still debated which side (pro-Europe or pro-Yanukovych) actually fired first.

Each side blames the other for that fatal shot. In the West, the prevailing view is that 'Mr. Yanukovych either permitted or ordered the use of sniper fire to terrorize the protesters into leaving the streets' ('Senate Foreign Relations Committee Hearing', 2015). The pro-Russian forces deny this accusation and insist that the fire was started by the pro-European demonstrators in order to initiate the regime change. Russian Foreign Minister Sergei Lavrov presented a 'huge amount of evidence', which, in his words, would disprove the claim that President Viktor Yanukovych was the first to order fire on protesters. In his turn, Lavrov accused the Ukrainian far-right forces of starting the fire in order to destabilize the situation ('Russia Rejects Kyiv's Sniper Claims; NATO Dismisses Moscow's Accusations', 2014).

The ill-fated sniper fire escalated the violence, and dozens of pro-Europe demonstrators were shot and killed on the streets in this culmination of a three-month long political crisis. Simultaneously, on the pro-government side, three police officers were killed and 21 officers were wounded (Uhler, 2015). February 20, 2014 became known in history as the February 2014 revolution and the day when the pro-Europe demonstrations turned violent, after weeks of peaceful protests. About 50 people died and about one hundred were wounded in just one day (Smith-Spark & Pleitgen, 2015). The casualties frightened Viktor Yanukovych, and on February 22 he fled to Russia. Before leaving, he agreed to a deal with the opposition movement, establishing 'a national unity government' and scheduling a presidential election for 2014 ('Putin Says He Told Officials Of Decision To Take Crimea In February 2014', 2015).

Yanukovych's departure signified a regime change in the politically divided country, in which the pro-European side was looking forward to ousting 'the notoriously corrupt president Viktor Yanukovych and his criminal cabal', as the Western media dubbed the Yanukovych administration (Kenarov, 2015). The U.S. government referred to Yanukovych's presidency as 'corrupt and increasingly authoritarian' ('Senate Foreign Relations Committee Hearing', 2015). The Russian media, on the contrary, blamed the United States for actively supporting Kiev's 'coup d'etat' (Rostovskiy, 2015). The fact that President Yanukovych fled to Russia was in and of itself telling. He was welcomed by the Russian government, as President Putin admitted in his speech in Sochi in October 2014. Quoted by the BBC, Vladimir Putin said: 'I will say it openly – he asked to be driven away [sic] to Russia, which we did' ('Putin: Russia Helped Yanukovych to Flee Ukraine', 2014). In the same speech, Putin lamented about the European and American influence in Ukraine.

As of 2015, Yanukovych is still in Russia under protection from the Russian government. It seems to have been a safe place for him, although speculations around the world arose when the international media reported in March 2015 that Yanukovych's son Victor, who was also in Russia, died in a traffic accident. His vehicle plunged through thin ice on Lake Baikal where he vacationed ('Son of Former Ukrainian Leader Viktor Yanukovych Dies in Accident in Russia', 2015). Even though his death was reported by the media around the world, it is possible that the interpretation of the events was not accurate.

The new Ukrainian government

With President Yanukovych gone, nobody else was standing in the way of a Europe-oriented Ukraine, and the hopes of more protected human rights and Westernized culture started to arise in the country. The new Ukrainian president, Petro Poroshenko, who won elections in 2014, has been extremely critical of his predecessor. Poroshenko's political career has risen on the pro-European integration agenda, based on minimizing economic ties with Russia. In his speech marking a one-year anniversary of the February revolution, Poroshenko said: 'Now, it is finally clear that we struggled on Maidan not against Yanukovych. He was just a cruel and obedient marionette'. He was even more critical of Russia's role in the events: 'Moscow was preparing for the liquidation and tearing Ukraine apart long before the victory of Maidan. They were expecting the fall of Yanukovych and accelerated the course of events' (Smith-Spark & Pleitgen, 2015).

Although the opposition to Russia was one of the main political themes in Petro Poroshenko's rise to power, his business interests had been connected to Russia for a long time. Before becoming president, he was a billionaire magnate, known as the 'Chocolate King' in Ukraine, and the products of his confectionery factory, Roshen, were very popular in Russia before the Ukrainian February revolution. After the Euromaidan protests, however, his products were banned by Russia's Rospotrebnadzor (a consumer protection agency). Russia's Chief Sanitation Doctor Gennady Onishchenko announced that Roshen confectionery products, manufactured in Ukrainian cities – Kiev, Mariupol, Kremenchuk, and Vinnytsia – were tested by Rospotrebnadzor, and the tests brought the following results:

> Unfortunately, our apprehensions have proven true, a fact which we sincerely regret. The identified irregularities give us reason to consider banning the imports of all products of this company… The quality and safety requirements were violated. There are reasons to speak of systemic violations of this country's legislation on the protection of consumers' rights.
>
> ('Confectionery; Russia Bans Imports from Ukraine's Roshen', 2013)

These violations of the product quality regulations were identified only in the Roshen products manufactured in Ukraine. There were no restrictions and statements about the Roshen products manufactured by the same company's factories in Lipetsk (Russia) and Klaipeda (Lithuania).

In response, the Roshen company issued a statement that it had 'always observed all the rules for the certification of confectionery products and fulfilled all sanitation norms by observing the provisions of statutory documents of Ukraine and the Russian Federation' ('Confectionery; Russia Bans Imports from Ukraine's Roshen', 2013). As of 2015, Roshen has been forced to withdraw from the Russian market after the company lost its case in the Russian appeal court. According to the 2015 Ukraine Food and Drink Report, the company's assets of $80 million were seized by a Russian court decision ('Ukraine Food & Drink Report – Q1 2015', 2015, p. 7).

Petro Poroshenko's election platform was based on denouncing corruption, distancing politically from Russia, and integrating with the European Union. His main challenger in the elections was former Prime Minister Yulia Tymoshenko who had previously signed multiple deals with Russia and could not claim full independence from Russia in her political past ('Ukraine: Profile: Petro Poroshenko, Ukraine's Presidential Candidate, "Chocolate King"', 2014).

Coming to power, Petro Poroshenko tried to limit the influence of the previous administration in the new political structure. After the elections, he famously said: 'It has born the new people. But unfortunately it left old politicians and I tried to do my best that the politicians, the government, the authorities and the president be adequate to the new people from the revolution' ('Ukraine: Profile: Petro Poroshenko, Ukraine's Presidential Candidate, "Chocolate King"', 2014). The new Ukrainian government made sure that the supporters of the former President Yanukovych did not stay in power and did not have access to government positions. In October 2014, President Poroshenko signed the law entitled 'On the Purification of Government', that defined legal and organizational grounds for the examination of civil servants and persons equated to them, as well as local government officials 'in order to restore trust in the government and create conditions for the construction of a new system of governmental bodies under the European standards' (CSIS, 2015).

The 'On the Purification of Government' law prevents public employees who worked for the Yanukovych administration for more than a year from taking government positions in the current administration (CSIS, 2015). About 80 per cent of the Ukrainian population supported this policy of lustration, under which all candidates for government and judicial positions are screened for their ties with the previous administration of President Yanukovych, as well as with the Soviet-period KGB. This law was drafted by the far-right political party 'Svoboda', and Euromaidan Council member Egor Sobolev, a prominent advocate of the policy, was appointed as Director of the Lustration Committee. Sobolev was quoted as saying that 'agents of foreign secret service, people who have been involved in corruption, and people who have been involved in the destruction of democracy' should be banned from the new Ukrainian government (Popova & Post, 2014).

Simultaneously, in October 2014, Ukraine passed anti-corruption legislation, establishing the National Anti-Corruption Bureau and demanding that public employees declare their assets. This legislation was supposed to bring Ukraine in line with the EU transparency laws (CSIS, 2015). In addition to the National Anti-Corruption Bureau, the Ukrainian government established the National Agency of Ukraine for Battling Corruption. Both agencies' directors were appointed through a very careful selection process, 'making procedure of appointing a director very rigorous, and firing [a director] – very difficult…' ('Rada Establishes Procedures for Appointing and Dismissing Director of Anti-Corruption Bureau', 2015).

By the end of 2014, the political structure of the new Ukraine was mostly set up. After Petro Poroshenko came to power, his most powerful political allies were Prime Minister Arseniy Yatsenyuk, and Dnipropetrovsk Governor Ihor Kolomoiskiy (Jarabik, 2015). Another influential opposition leader, Vitali Klitschko, became Mayor of Kiev (Kenarov, 2015). All of them had strong pro-Europe views.

The war of words between Russia and Ukraine

The conflict between Ukraine and Russia has been fueled by both governments' use of strong rhetoric, especially references to WWII. For example, the political activities of the All-Ukrainian Union Svoboda ('Freedom') and the Pravyi Sektor ('Right Sector') – the parties that had played a crucial role in the February revolution and remained on the front lines of the new Ukrainian order – have received the opposite descriptions and evaluations. On the Ukrainian side, Svoboda and the Right Sector have been highly influential and respected, as they led the Maidan Square uprising (Bilozerska, 2015). On the Russian side, however, both parties have often been described as having 'neo-Nazi orientation'.

Since the parliamentary elections in October 2012, Svoboda has been considered the most powerful far-right political party in Ukraine, gaining 10 per cent of the national vote and the majority of votes in the Lviv city council (Lviv is the most pro-Western city in Ukraine). The party has a long history: it was initially established in 1995 as the Social-National Party of Ukraine (SNPU) by Oleh Tyahnybok, who in 1998 was elected to the Verkhovna Rada (the Ukrainian Parliament). In 2002, SNPU joined the Our Ukraine Bloc of Victor Yushchenko, and two years later, SNPU changed its name into Svoboda and strongly supported the Orange Revolution of 2004. Most recently, after the Ukrainian February revolution of 2014, members of Svoboda became key members of the government: several ministers, a deputy prime minister, and the prosecutor general (Mudde, 2014).

Similar to Svoboda, the Right Sector members have held several important government positions, including deputy secretary of the National Security and Defense Council, even though unlike Svoboda, which has a well-defined political structure, the Right Sector has often been described as a 'coalition of mostly smaller far right groups'. The self-proclaimed nationalist coalition's mission has been 'defending the values of white, Christian Europe against the loss of the nation [Ukraine] and deregionalisation' (Mudde, 2014).

That is not how Svoboda and the Right Sector have been perceived on the Russian side: the Russian media and government's view of these parties has been very different from the official Ukrainian version. In Russia, the Right Sector and Svoboda have often been described as radical militants and ultra-nationalists ('Ukraina: Vostok Vzbuntovalsya', 2014). The Russian media has also referred to Kiev's far-right parties as 'banderovtsy', which is a very offensive term in Russia (Oberemenko, 2014). The term 'banderovtsy' refers to Stepan Bandera (1909–1959), a very controversial historical figure who fought against the Soviet government during WWII. Calling the new Ukrainian government 'banderovtsy' is indicative of Russia's view of these far-right parties as neo-Nazi (Oberemenko, 2014).

In April 2015, main Russian state-controlled TV channels – NTV, Rossiya 1, and Channel One – reported on an arrival of U.S. military instructors to Ukraine to train the Ukrainian military, including what the Russian media called the 'neo-Nazi battalions', such as the Azov batallion ('Russian TV News: US Military Training for Ukrainian "Neo-Nazis"', 2015). The U.S.–Ukraine military cooperation has been evaluated by the Russian media as 'non-stop NATO exercises being held at the initiative of the USA'. Referring to the long-term grievance about NATO's expansion, the Russian TV channel NTV showed the map of the increasing number of American military bases near Russia's borders ('Russian TV News: US Military Training for Ukrainian "Neo-Nazis"', 2015).

In addition to 'banderovtsy', the Russian media used another very offensive Russian term while talking about the Ukrainian military – 'karateli'. The term 'karateli' (punishers) has a very strong connotation, meaning Nazi troops who killed the Russian military and civilians during the Great Patriotic War (WWII fought on the Russian territory). Currently, the term is used by the Russian media to describe the Ukrainian army ('Russian TV News: US Military Training for Ukrainian "Neo-Nazis"', 2015). On the Voice of Russia news channel, the conflict with Ukraine has been referred to as 'the West and its agents have unleashed Nazi and fascist forces bent on the destruction of Russians, the "Moskali" and anyone they can demonize as "pro-Russian"'. Talking about the celebration of the 70th anniversary of the D-Day in Europe, the Russian channel lamented that the West was 'supporting Nazis in Kiev and ignoring war crimes being committed by fascists in Ukraine' (Robles, 2014). In the same media report, the current Ukrainian government was referred to as 'the fascist junta' (Robles, 2014). The Ukrainian government, however, has denied that the Right Sector and Svoboda had a pro-Nazi orientation. On the contrary, the Ukrainian media has often referred to Russia's policies toward Ukraine as 'fascist' (Goble, 2015).

Calling someone a 'banderovits' in Russia is an extreme offense and an expression of contempt. According to the Russian sources, Bandera was a Nazi collaborator during WWII, who received 2.5 million German marks from the Nazis for reconnaissance and terrorism activities in the USSR. He urged the Ukrainian people to help the German army and fight the Moscow-imposed bolshevism. While being connected to the Nazis, Bandera tried to declare an independent Ukrainian state and to create a Ukrainian government, without consulting with Germany. For that attempt he was jailed by the Nazis, and his fellow commanders were executed. After spending more than a year in jail, Bandera was released and continued to collaborate with Nazi Germany: his legion performed police functions on Nazi-occupied territory. Bandera was leading the Organization of Ukrainian Nationalists, headquartered in West Germany, until his death in 1959 ('Biografia Ideologa Ukrainskogo Natsionalizma Stepana Bandery', 2015).

The term 'banderovits', however, does not have a negative connotation in the pro-Europe Ukraine. The Ukrainian government has offered an opposite description of Bandera's life: pro-Europe politicians consider Stepan Bandera a Ukrainian hero. Long before the February revolution of 2014, the Ukrainian President Viktor Yuschenko presented a state order to Bandera, making Bandera an official hero for his activities as the leader of the Organization of Ukrainian Nationalists (OUN). During the ceremony, Viktor Yuschenko said: 'I decree to confer the rank of a Hero of Ukraine to Stepan Bandera and present a state order to him' for 'defending national ideas and battling for an independent Ukrainian state' ('Stepan Bandera Becomes Ukrainian Hero', 2010). In 2015, the far-right Ukrainian groups have celebrated Bandera's anniversary and publicly praised him for fighting the 'Russian and Polish occupation in Ukraine' ('Far-right Ukrainians Mark Anniversary of Nationalist Hero Stepan Bandera', 2015).

A year before that, in 2014, another law was supposed to recognize the Ukrainian Insurgent Army as national heroes (this law was narrowly defeated in the Ukrainian parliament). The Ukrainian Insurgent Army was another controversial group that collaborated with Nazi Germany during WWII (CSIS, 2015). On October 14, 2014, one of the main Russian TV channels – Channel

One – reported from Kiev about the anniversary celebration of the Ukrainian Insurgent Army. The Russian media described it as the 'neo-Nazi celebration that adds to the racial and ethnic rifts between countries' appeals to violence against Russian citizens' (Brainin, 2014).

In the Russian media and in public opinion, Ukraine's glorifying of those controversial historical figures and movements has been considered as siding with the Nazi regime, which is the worst offense in Russian culture. As a result, Ukraine's praising these historical figures and groups evoked an extremely negative response in Russia, both at the government level and in the popular sentiment, as WWII is an event about which the Russian population feels very strongly. Ukrainian Prime Minister Arseniy Yatsenyuk, for example, has been presented in the Russian media as 'pandering to all of his neo-Nazi supporters fighting for his regime in eastern Ukraine' (Uhler, 2015).

The war of words did not stop at the interpretation of Stepan Bandera's life – the whole of WWII has been reinterpreted on both sides. The views now differ even on who actually started the Great Patriotic War, which is WWII fought on Soviet land. Ukrainian Prime Minister Arseniy Yatsenyuk, for example, in his interview on German television on January 8, 2015 said: 'I will not allow the Russians to march across Ukraine and Germany, as they did in WWII…We all very well remember the Soviet invasion of Ukraine and Germany, and we have to avoid it' (Uhler, 2015). In Russia, Yatsenyuk's statement was read as a claim that Russia had invaded Germany and Ukraine in 1941. This is a complete opposite from the conventional knowledge in Russia that Hitler invaded the USSR (which included both the Russian Federative Republic and the Ukrainian Federative Republic) in June 1941. Not surprisingly, Yatsenyuk's statement created an outrage in Russia.

Not only has the beginning of the war been questioned, but also the end of it – Victory Day (May 9, 1945). In Russia this day is celebrated as the day when the Great Patriotic War ended with the Soviet victory, and it has been a sacred holiday in Russia, widely celebrated by the Russian population. The last celebration, on May 9, 2015, was accompanied by a 16,000-troop parade on Moscow's Red Square. At the parade, Vladimir Putin addressed the nation, talking about how 'our grandfathers lived through horrible suffering'. His speech resonated with the Russian people, because almost every family had lost someone in that war. It was the 70th anniversary of Victory Day, and hundreds of thousands of people around Russia took to the streets, carrying portraits of their family members who had lost their lives in the war. However, indicative of the current state of Russia–West relations, Western leaders did not attend the parade. Instead, Vladimir Putin was joined on Red Square by Chinese President Xi Jinping, Cuban leader Raul Castro, and Zimbabwean President Robert Mugabe ('Russia Marks Victory Over Nazi Germany With Military Parade', 2015).

In post-Soviet Ukraine, however, this holiday has received a different interpretation. Since Ukrainians 'fought on both sides', and the Ukrainian Partisan Army 'was loosely allied with Nazi Germany', the Ukrainian far-right groups have long considered Victory Day as a 'fraught' holiday (MacFarquhar,

2014). The pro-Russian Eastern Ukraine, however, has followed the Russian tradition of celebrating May 9th. Eastern Ukraine's celebrating of Victory Day in 2014 and 2015 has been an act of defiance under the new political conditions in the post-February revolution Ukraine. In 2014, the New York Times reporters interviewed people in Donetsk before the holiday, and a typical response was that 'Kiev doesn't want us to celebrate this holiday because they are supported by neo-Nazis' (MacFarquhar, 2014).

In addition to the harsh rhetoric and the revision of history, another complicating factor in Russia–Ukraine relations has been actual actions and policies, proposed and implemented by the far-right political groups in Ukraine. Some of these policies were used by Vladimir Putin and Russia as the grounds to safeguard the Russian-speaking population in Ukraine.

One such aspect has been the visual symbols and statements of the Right Sector and Svoboda, which made even some experts in the West concerned about the extreme far-right nature of these parties (Uhler, 2015). For example, a leading American scholar of Russian politics, Robert English, noted, referring to the Right Sector and Svoboda:

> These are groups whose thuggish young legions still sport a swastika-like symbol, whose leaders have publicly praised many aspects of Nazism and who venerate the World War II nationalist leader Stepan Bandera, whose troops occasionally collaborated with Hitler's and massacred thousands of Poles and Jews.
>
> (English, 2013)

When the far-right came to power in Ukraine in February 2014, they started to implement changes in the status of the Russian language. For a long time, the language issue has been by far the most debated and divisive in modern Ukraine. During the Yanukovych administration, the Russian language had a privileged status: on July 3, 2012, Viktor Yanukovych introduced a law entitled 'On Principles of the State Language Policy'. The law strengthened the position of the Russian and other national minority languages in those Ukrainian regions in which national minorities constitute more than 10 per cent of the population (i.e. Crimea and Eastern Ukraine) ('Language Law Splits Ukrainian Society – OSCE High Commissioner', 2012).

The 2012 law became a very divisive and irritating factor for pro-European Ukrainian forces, because in it they saw an attempt to diminish the importance of the Ukrainian language, which has been regarded by the pro-European movement as an 'ethno-consolidating factor'. In their view, the 2012 law 'undercut the status of Ukrainian as the sole state language' (Dibrova, 2013). Not surprisingly, one of the first policies that the Right Sector and Svoboda proposed after coming to power in 2014 was to 'correct' the situation and to strengthen the role of the Ukrainian language. Svoboda publicly 'called to abolish the autonomy that protects Crimea's Russian heritage'. The far-right Ukrainian parties called for the exclusion of the Russian language from public schools and tried to implement the requirement that only Ukrainian-speaking people be provided with Ukrainian citizenship (English, 2013). Strengthening the status of the Ukrainian language was presented as an instrument to increase Ukrainian national unity.

Predictably, these language initiatives by the Right Sector raised extreme concerns in Crimea and the Eastern part of Ukraine, in which the majority of the population spoke Russian as their native language. They perceived the Ukrainian language as an encroachment on their right to continue using Russian in all aspects of life. As a result of those concerns, the Russian government announced in official statements that the Russian-speaking population in Ukraine needed to be protected from potential discrimination, by all possible means, including using the Russian military contingent in the Crimean peninsula (predominantly Russian-speaking). This protection resulted eventually in a referendum in Crimea about the peninsula's self-determination. Russian politicians stated that even if Eastern Ukraine expressed a desire to join Russia, the Russian government would 'have to agree' ('Ukraina: Vostok Vzbuntovalsya', 2014).

Crimea joins Russia

Crimea was one region in which the issues with language status (Russian vs. Ukrainian) and the debate about the far-right parties played a crucial role. Before the conflict, about 60 per cent of Crimea's population was Russian, about one quarter was Ukrainian, and about 12 per cent was Tatar ('Ukraine Crisis: Crimea's Troubled History is Dictating Modern Day Events', 2014).

Crimea's history was similar to other post-Soviet ethnic enclaves, in which the borders were redrawn during the Soviet time. The Russian Empire annexed Crimea in 1783, taking the peninsula from the Ottoman Empire. Since then it has been a strategically important post for Russia, hosting the military bases and the Russian fleet (M. Kramer, 2014).

During the Soviet time, Crimea stayed a part of the Russian Soviet Federative Socialist Republic (RSFSR) until 1954, when the Communist Party Secretary General Nikita Khrushchev 'gifted' Crimea to the Ukrainian Soviet Socialist Republic – his native land ('Putin Says He Told Officials Of Decision To Take Crimea In February 2014', 2015). At the time of the transfer, the population of Crimea was even more predominantly Russian than it is today – about 75 per cent was ethnically Russian (M. Kramer, 2014). Since both Soviet republics were part of the USSR, it did not matter that much that Crimea became part of Ukraine. After the dissolution of the Soviet Union, however, a significant portion of the Russian population resented the fact that Crimea, being such an important port and a tourist spot, remained part of the post-Soviet Ukraine. Others were simply puzzled with Khrushchev's decision.

In 1954, the Soviet government did not deem it necessary to explain to the Soviet people the decision to transfer Crimea to the Ukrainian Federative Republic. Only after the breakup of the Soviet Union, when some archives became open, modern historians started to look into this issue. Khrushchev had been a political commissar and the head of the Communist Party in Ukraine in the 1930s and 1940s. Among other responsibilities, he was a commander of the Soviet military forces in a civil war in Volynia and Galicia, which were some of the last Ukrainian regions to be annexed by the Soviet Union. The Soviet Army had to use brutal methods in order to annex those regions, and historians have found that by transferring Crimea to Ukraine in 1954, Khrushchev was trying to amend the memory of violence and to prove that the 'friendship' between Russian and Ukrainian people was genuine and deep. By this action, the Soviet leader also strengthened his own political power, as he was challenged by Soviet Prime Minister Georgii Malenkov. Even though Khrushchev acted in accordance with the Soviet Constitution at the time and both the RSFSR and UrkSSR gave their consent to the transfer, this political act had huge reverberation in post-Soviet Russia (M. Kramer, 2014).

When the Soviet Union disintegrated, some of the Soviet nuclear arsenal was located on Ukrainian territory. Ukraine reluctantly gave up the nuclear weapons and returned them to Russia. In response, the international community, and Russia especially, guaranteed the Ukrainian sovereignty ('Ukraine Ratifies the Missile Pact, But Delays Ending Nuclear Status', 1993). It was a bad decision for Ukraine, which was the third largest nuclear power in the world. Under those conditions, the possibility of losing Crimea was for Ukraine a breach of that promise. The Ukrainian government even compared losing Crimea to 'Hitler's occupation of Sudetenland and the Soviet Union's invasion of Hungary in 1956 and Czechoslovakia in 1968' (Onyschuk, 2014).

Nevertheless, in response to the Ukrainian language initiatives and other pro-European policies in 2014 by the Right Sector and Svoboda, Crimea put forth a Russia-backed referendum in order to determine if the population of Crimea would choose to join Russia. The official purpose of the referendum, held on March 16, 2014, was to form the Crimean Federal District, making it a part of Russia (Kenarov, 2015). The far-right parties' actions in Ukraine after the 2014 February Ukrainian revolution made it easier for the government of Crimea to achieve the consensus, backed by Russia, and for Vladimir Putin to explain the need to protect Crimea to his domestic audiences (Smith-Spark & Pleitgen, 2015). In his public remarks, Vladimir Putin said:

> I think 2014 will also be an important year in the annals of Sevastopol and our whole country, as the year when people living here firmly decided to be together with Russia, and thus confirmed their faith in the historic truth and the memory of our forefathers... There is a lot of work ahead, but we will overcome all the difficulties because we are together, and that means we have become even stronger.
>
> (MacFarquhar, 2014)

An interesting fact is that the events in Ukraine started shortly after the 2014 Winter Olympic Games in Sochi, held on February 7–23, 2015. The games have been the most expensive in the Olympic history (of the planned cost of $41 bn), bringing a profit of only $54 m. For comparison, the London Olympic Games in 2012 cost only $19 bn (Weir, 2013). Despite some initial hiccups, such as unfinished hotels and brown tap water (widely reported by the media around the world), the games were a success, and the International Olympic Committee President Thomas Bach praised the games as 'seamless' and 'excellent' ('Russia: Sochi Olympic Games Made USD 53-M profit – IOC', 2015). Vladimir Putin monitored the Olympic project implementation very closely, and almost the whole nation took a great pride in it. In addition to the games themselves, Putin planned to use Sochi facilities for other events, such as Formula-1 and a G-8 Summit (Weir, 2013).

Because of high costs of the games, however, not everyone in Russia was happy with the Olympic construction. The final cost of the games was $51 bn, which is even higher than the games in China with the price tag of $40 bn (Vasilyeva, 2015). Critics pointed out that the games were one of the 'showcase' projects, similar to the 2012 Asia-Pacific Economic Cooperation Summit in Vladivostok, in which billions of dollars spent did not sufficiently benefit an average Russian, as the country needed serious updates of the infrastructure (roads, schools, etc.) (Weir, 2013). For example, the international media reported that the problems of unreliable electricity supply and poor transportation in Sochi were not resolved even after the Olympics. In the meantime, Russia is scheduled to host another major global event – the 2018 World Cup – and a similar debate has been happening about this grandiose construction (Vasilyeva, 2015). As the ruble weakened in 2014 because of the falling oil prices, the price tag in rubles for building the 2018 World Cup facilities has been predicted to rise by 40 per cent, according to Russia's Sports Minister Vitaliy Mutko. In order to minimize the losses, the Ministry decided to build the World Cup facilities (12 stadiums), mostly using Russian materials ('Russia: Ruble's Fall Sends Cost of World Cup Stadiums Soaring 40 Per Cent', 2015).

On the international arena, the Olympic Games were supposed to bring good will toward Russia and to improve peaceful relations with other countries. The media called the games a rebranding exercise for Russia. What happened in reality was that the Sochi Olympic Games were followed by the events in Ukraine, and Russia's relations with the West drastically worsened. The Crimea referendum, which produced the vote to join Russia, happened even before the Paralympic Games ended (Harris, 2014).

In Russia, the majority of the population met the Olympic Games and the Crimea referendum with elation and overwhelming approval. Vladimir Putin's approval rating, according to the Russian Public Opinion Research Center (VTsIOM), reached 67 per cent. Even opposition leaders, such as Ksenia Sobchak, acknowledged Putin's popularity among the population (Sonne, 2014).

Historically, the coincidence of the two events – the Olympic Games and Crimea's joining Russia – was eerily similar to the Russia–Georgia war that started in South Ossetia during the opening ceremony of the 2008 Beijing Olympic Games (Panzeri, 2008). The similarities between the two events (of 2008 and of 2014) did not add popularity in the West for Russia.

As it might be expected, the Western nations, as well as most other countries, did not recognize the results of the referendum as valid, stating that the referendum had been orchestrated by Russia in violation of the Ukrainian sovereignty and international law. U.S. Ambassador to the United Nations Samantha Power said: 'The whole world knows that the referendum… in Crimea was hatched in the Kremlin and midwifed by the Russian military' (Magee, 2014). She called the referendum 'illegitimate and without any legal effect' and added that the referendum was 'inconsistent with Ukraine's constitution and international law' (Magee, 2014).

The United States drafted a United Nations resolution that was supposed to announce the referendum as invalid. Russia, predictably, vetoed the resolution, and China abstained from the vote. Russia responded to the resolution by referring to previous historic cases of self-determination: 'It is also understandable that enjoyment of the right of self-determination as to separation from an existing state is an extraordinary measure, applied when future coexistence within a single state becomes impossible' (Magee, 2014). Vladimir Putin said that it was impossible for Crimea to stay within Ukraine because of the 'unconstitutional coup' carried out by 'radicals' in February 2014 (Magee, 2014).

The Russian government has emphasized that the Crimean referendum is no different from other referenda that were accepted by the international community, such as in Scotland and Catalonia ('Referendum on Status of Crimea Fits General European Tendencies – Markov', 2014). Expressing his dissatisfaction with the West's reaction to the Crimea referendum, Vladimir Putin also compared Crimea with Kosovo. In Putin's words, there was no violation of international law by conducting the Crimea referendum: 'Kosovo… declared its independence by parliamentary decision alone. In Crimea, people did not just make a parliamentary decision, they held a referendum, and its results were simply stunning' ('Putin: What was Done in Crimea was Not in Any Way Different from What Had Been Done in Kosovo', 2014). Putin emphasized that the people of Kosovo had not had to ask permission for their independence from the 'central authorities of the state' where they were living at the time, so Crimea also did not need to ask permission of the Ukrainian government.

Russia has not denied the support it had provided to Crimea around the referendum. Vladimir Putin publicly admitted that the decision to annex Crimea had been made shortly after Viktor Yanukovych had lost control of the protests on the Maidan Square in February 2014. According to Putin, the decision was made in a meeting at the Kremlin on February 23, 2014. It was the same meeting in which Putin was ordered 'to save the Ukrainian president's life', by helping Yanukovych flee to Russia. In his televised remarks, Putin confirmed that Russian soldiers had been dispatched to Crimea, with an official mission to protect Russian-speaking people ('Putin Says He Told Officials Of Decision To Take Crimea In February 2014', 2015).

Speaking about the Russian military contingent in Crimea before the referendum, Vladimir Putin insisted that the official purpose of the troops was to protect the Russian-speaking population:

> … I did not hide the fact that our goal was to ensure proper conditions for the people of Crimea to be able to freely express their will. And so we had to take the necessary measures in order to prevent the situation in Crimea unfolding the way it is now unfolding in southeastern Ukraine…Of course, the Russian servicemen did back the Crimean self-defense forces.
>
> ('Direct Line with Vladimir Putin on April 17, 2014', 2014)

President Putin also mentioned in his interviews that the Russian forces had entered Crimea around the time of the March 2014 referendum:

> Yes, I make no secret of it, it is a fact and we never concealed that our Armed Forces, let us be clear, blocked Ukrainian armed forces stationed in Crimea, not to force anybody to vote, which is impossible, but to avoid bloodshed, to give the people an opportunity to express their own opinion about how they want to shape their future and the future of their children.
>
> ('Putin: What Was Done in Crimea Was Not in Any Way Different from What Had Been Done in Kosovo', 2014)

Since then, Russia has even more increased its military presence in Crimea by locating several strategic TU-22M3 bombers there (Dubrovskaya, 2015).

In response to the Crimea annexation and Russia's involvement in the process, most countries issued statements of disapproval, which was especially harsh from the United States. The U.S. government has repeatedly issued strong statements of condemnation and urged Vladimir Putin to end the 'occupation of Crimea'. The spokesperson for the U.S. Department of State called the Crimea vote in March 2014 a 'sham' referendum in 'clear violation of Ukrainian law and the Ukrainian constitution' (Psaki, 2015b). The Department of State presented statistics showing that the situation in Crimea worsened since the 'liberation', because of violations of basic human rights, such as freedom of speech and political freedoms. The statement ended with a promise that the United States would continue 'to support Ukraine's sovereignty, territorial integrity, and right to self-determination', and economic sanctions for Russia (Psaki, 2015b).

Furthermore, the U.S. Department of State issued another statement, connecting the Crimea referendum to the situation in South Ossetia and Abkhazia. This last statement strongly asserted that the United States supported Georgia's territorial integrity and did not recognize the independence of the two breakaway republics. The spokesperson denounced treaties that Russia had signed in 2014–2015 with South Ossetia and Abkhazia as having no international legitimacy (Psaki, 2015a).

In response to the U.S. government's statements, Vladimir Putin's Press Secretary Dmitry Peskov asked Washington not to 'interfere in the internal affairs of Russia'. Peskov stressed that Crimea was a Russian region, and Russia was not planning to consult anyone on Crimean affairs. He emphasized that Crimea 'was not occupied', but that it was annexed as a result of the people's referendum and good will (Dubrovskaya, 2015).

In addition to the international backlash, the Russian government has had a more urgent task on its hands: to prevent an economic collapse in Crimea. After Crimea became a part of Russia on March 21, 2014, less than a week after the 97 per cent approval vote in the Crimea referendum, the Russian government started to work on supplying Crimea with fuel sources and electricity ('Putin Says He Told Officials Of Decision To Take Crimea In February 2014', 2015). Such supplies were hard to provide without Ukraine's cooperation. One such agreement between Russia and Ukraine in early 2015 was about shipments of 1,500 MW of electricity from Russia to Ukraine and then to Crimea. Russia's Prime Minister Dmitry Medvedev described this agreement as 'beneficial to both parties', because cheap Russian electricity would also help support the Ukrainian economy and energy systems (Interfax, 2015).

In connection to those supplies of electricity and fuels, as well as other forms of economic aid to Crimea, Russian commentators have discussed the 'price' of Crimea's annexation for the Russian federal budget. To modernize the outdated Crimean infrastructure, Russia would have to invest at least 240 billion rubles (around $8 billion at the exchange rate in April 2014, when this forecast was discussed). Some 200 billion rubles (about $6.5 billion) more would be needed for economic assistance to improve living standards in Crimea and to compensate the residents for higher electricity and water prices. Currently, Crimea is fully dependent on Ukraine for water supply ('Ukraina: Vostok Vzbuntovalsya', 2014).

Despite those expenses, the Russian officials and media argue that the annexation will bring multiple benefits for Russia. First, Crimea has an extensive manufacturing base, such as chemical plants and other manufacturing industries, including wine production, which Russia plans to utilize. Second, Crimea is a major tourism destination spot for Russia, which has few warm-weather cities. Third, Crimea possesses large reserves of shale oil and natural gas. Potentially, 5–6 billion cubic meters of gas and 1 million tons of oil could be extracted there every year.

Fourth, after the dissolution of the Soviet Union, Russia had to pay to Ukraine $100 million per year for renting the port of Sevastopol for the Russian Black Sea Fleet ('Ukraina: Vostok Vzbuntovalsya', 2014). Russia was renting the port under the 1995 agreement between Russian President Boris Yeltsin and Ukrainian President Leonid Kuchma, which stipulated that Russia receive 82 per cent of the former Soviet Black Sea Fleet, with the remaining vessels taken by Ukraine. Russia's rent was mostly paid in 'the form of energy supplies and debt forgiveness' (Erlanger, 1995). The Black Sea Fleet is a matter of national pride for Russia, and the majority of the Russian population was not happy about the fleet's location on Ukrainian territory. With the annexation of Crimea, this 'injustice' was corrected. For all of the reasons above, the majority in Russia supported Crimea's return, even though the Russian opposition movement was very critical about it ('Putin Delivers Crimea Annexation Address to Russian Parliament', 2014).

Russian opposition commentators have explored how the Crimean economy has been developing since the annexation. By the first anniversary of the annexation, the critics provided evidence that the gross regional product fell by

11 per cent in one year, compared to the 8 per cent fall in the rest of Ukraine. The number of tourists in Crimea fell from the annual average of 6 million to 4 million. A sharp fall in other economic indicators (e.g., incomes, employment) had a destabilizing effect (Oreshkin, 2015). At the same time, prices for consumer goods increased by 42 per cent, making life unaffordable for many Crimeans (Malysheva, 2015).

To respond to the critics, the Russian government gave its own account of the Crimean economy on the first anniversary of Crimea's annexation. Russia's Minister of Crimean Affairs Oleg Saveliev blamed the Ukrainian and Western sanctions for Crimea's economic troubles. Ukrainian investments stopped coming to Crimean businesses, and the peninsula was cut off from mainland Ukraine. One of the main transportation routes available for Crimea has been by ferry to the port of Feodosia. To improve the situation, Russia expanded the Kerch ferry system and the airport capacity. In financial assistance to Crimea, in 2014 alone, Russia spent 124 billion rubles, which is from $4 billion to $2 billion, depending on the fluctuating exchange rate (from 30 to 70 rubles per dollar) (Malysheva, 2015). Despite large amounts of Russian assistance to Crimea, which could alternatively have been used for infrastructure improvement in mainland Russia, the prevailing sentiment in Russian society has been that gaining Crimea back is an event above the simple arithmetic of gains and losses. It has been considered a 'God-given' event and a bestowed gift returned back to Russia (Oberemenko, 2014).

The Eastern Ukraine secession movement

The economic, political, and linguistic disagreements between Russia and Ukraine during the post-Cold War years have led to the lowest point in 2014, when the conflict in Eastern Ukraine erupted in casualties and the flow of refugees. The Eastern Ukrainian separatist movement has been perceived in Russia as activists who were trying to liberate Donetsk, Kharkov, and Luhansk from the Right Sector militants (Uhler, 2015). The secessionist Donbass region comprises about one tenth of Ukraine's territory and includes the cities of Luhansk and Donetsk, dubbed as the Donetsk People's Republics in March 2014. In total, one year of fighting in this conflict caused more than 5,000 casualties ('Nine Soldiers Killed in 24 Hours as Violence Surges in Ukraine', 2015). The Ukrainian government has repeatedly accused Russia of backing separatists and providing them with military and economic assistance (Magee, 2014). The Russian government has consistently denied this.

The conflict started to escalate when the pro-Russian regions of Donetsk and Luhansk declared independence in May 2014, after a referendum was held in those regions: 'We, the people of the Donetsk People's Republic, based on the results of the May 11, 2014, referendum declare that henceforth the Donetsk People's Republic will be deemed a sovereign state' (Leonard & Isachenkov, 2014). The statement was based on the referendum results, in which 89 per cent of Donetsk voters and 96 per cent of Luhansk voters voted for sovereignty, with the turnout exceeding 70 per cent. Moreover, the co-chairman of the Donetsk newly formed government, Denis Pushilin, announced that Donetsk would like to join Russia, because '[t]he people of Donetsk have always been part of the Russian world, regardless of ethnic affiliation' (Leonard & Isachenkov, 2014).

At that point, Ukrainian Prime Minister Arseniy Yatsenyuk was still hoping to 'launch the broad national dialogue with the east, center, the west, and all of Ukraine' (Leonard & Isachenkov, 2014). The Russian government also did not seem to be willing to annex the Eastern Ukraine regions in the same fashion as Crimea had been annexed. Instead, Vladimir Putin stated that '[t]she practical implementation of the referendum results should proceed in a civilized way without any throwbacks to violence through a dialogue between representatives of Kyiv [sic], Donetsk and Luhansk' (Leonard & Isachenkov, 2014).

However, the tensions between the self-proclaimed Eastern Ukrainian government and Kiev quickly turned into military actions. The violence actually started on the day of the referendum itself – May 11, 2014 – when armed men opened fire in Krasnoarmeisk. The Western media reported that the armed men identified themselves as members of the Ukrainian national guards (the reserves of the Armed Forces of Ukraine), but those claims were denied by the Ukrainian Interior Ministry (Leonard & Isachenkov, 2014).

This military conflict has several important facets to it: first, the whole nature of the conflict and the description of the fighting sides in the Western rhetoric and in the Russian media and official statements; second, the origins of the conflict, stemming from Russia–West disagreements on the role of Ukraine in the international system, NATO's expansion, and U.S.–Russia relations; third, the fighting itself and Russia's alleged involvement; finally, the economic aspects of Eastern Ukraine's position.

To start with, the literature and the media have yet to reach consensus about what to call the government of Eastern Ukraine. The Russian media emphasized that the Eastern Ukrainian population were not 'separatists' or 'Russian agents', but they were Ukrainian citizens that have the same right to decide the fate of Ukraine as do the Western Ukrainian population. The Russian media have often referred to Eastern Ukrainians as 'activists' ('Ukraina: Vostok Vzbuntovalsya', 2014). In Western media, Eastern Ukrainians have been most commonly referred to as 'separatists' (A. E. Kramer, 2014) or 'pro-Moscow insurgents' (Leonard & Isachenkov, 2014). In the Ukrainian official rhetoric, not surprisingly, the language is even stronger. The Ukrainian media has often referred to the Eastern Ukrainian fighters as 'terrorists' ('SBU Opens Criminal Case on Russian Bank Financing Terrorists in Eastern Ukraine', 2014). Even before the referendum in Donetsk and Luhansk on May 11, 2014, the Security Service of Ukraine (SBU) started a criminal case, in April 2015, against a Russian bank that was 'used to finance terrorist groups in east Ukraine… under the slogans of pro-Russian separatism' ('SBU Opens Criminal Case on Russian Bank Financing Terrorists in Eastern Ukraine', 2014). Those 'terrorist groups', according to SBU, 'committed violence against citizens, pogroms, arson, destruction of property, seizure of government buildings, resisted representatives of the authorities and law enforcement agencies with the use of weapons, as well as actions aimed at preparing to commit terrorist attacks' ('SBU Opens Criminal Case on Russian Bank Financing Terrorists in Eastern Ukraine', 2014). For the purpose of this study, the term 'separatists' is used as a definition of a 'person who supports the separation of a particular group of people from a larger body on the basis of ethnicity, religion, or gender' ('Separatist (definition)', 2015).

This conflict is rooted in major disagreements between Russia and the West (especially, the United States) about Ukraine's position in the post-Cold War order. Russia has long insisted that Ukraine be in its economic and political sphere of interest, strongly opposing Ukraine's admission into NATO military alliance. The West, however, sees Ukraine as a potential candidate for the European Union in the future, integrating Ukraine in the European economic system step by step.

In March 2015, an influential Russian newspaper *Moskovskiy Komsomolets* summed up Russia's official sentiment toward Ukraine. First, Ukraine must stay a 'non-aligned' country – not a member of NATO or the European Union. Second, NATO's expansion to the East needs to stop. The conflict with the West, the Russian government stated, may end only if these two conditions are satisfied by the West (Rostovskiy, 2015).

Some of the turmoil between Russia and Ukraine stems also from Ukraine's relations with the United States. Referring to the 2014 conflict in Ukraine, Russian economist Mikhail Khazin said that Ukraine became a 'hostage' of American strategic interests (Gudkova, 2014). He even called Viktor Yanukovych an 'agent' of the United States in Ukraine who, in return for U.S. guarantees to preserve his personal wealth, promised to sign a deal with the European Union, making Ukraine a cheap resource appendix of the European Union. Yanukovych's public refusal to sign the deal with the EU, Khazin argues, was just a façade, while Yanukovych was planning to build a deep-water port in Crimea jointly with China, using Chinese capital ($15 billion) and labor – and the cooperation with China was unacceptable for the United States (Gudkova, 2014). The Maidan Square demonstrations, Khazin continued, were also financed and trained by the United States.

According to the Russian media, the disagreements between the United States and Russia have been the foundation of the Russia–Ukraine conflict. The overall tone of coverage in the Russian media has been focused on the U.S. role in the conflict and beyond: in the global economy, the global military system, European energy security, and many other aspects. Even the public poll results, published in the most-read newspaper in Russia – 'Argumenti i Fakti' (*Arguments and Facts*) – show that the Russian population had a very negative image of the United States. According to a poll by 'Levada-Center' in 2013, in response to the question 'What role does the U.S. play in the modern world?' about 50 per cent of respondents replied 'in general, negative' in 2013. It was a 7 per cent jump from the same poll in 2012, regarding the same question ('S Amerikoi nado derzhat ukho vstro', 2013).

The media in Russia quoted examples of Western support to Ukraine as evidence of American and Western involvement in Ukraine. For example, the Russian media reported in April 2014 that the United States gave Ukraine the credit guarantees for $1 billion ('Ukraina: Vostok Vzbuntovalsya', 2014). The American ally, Germany, considered deployment of German soldiers in Ukraine in order to help stabilize the situation but instead decided in October 2014 to provide civilian aid, such as water, beds, generators, and survival kits (CSIS, 2015).

On the U.S. side, American experts have argued that Russia has been militarily involved in Eastern Ukraine. In his testimony to the U.S. Congress, former United States Ambassador to Ukraine John Herbst stated that Russia was waging a 'hybrid war' in Ukraine by using 'little green men' ('Senate Foreign Relations Committee Hearing', 2015). He also outlined the so-called Putin Doctrine:

> [It] calls for a Russian sphere of influence in the former Soviet space; describes Georgia, Ukraine, and now Kazakhstan as failed or artificial states; asserts Moscow's right and even duty to protect not just ethnic Russians, but Russian speakers wherever they happen to reside; and calls for new rules for the post-Cold War order, or 'there will be no rules'.
>
> ('Senate Foreign Relations Committee Hearing', 2015)

The phrase 'little green men' became a catchy description of the alleged Russian troops in Ukraine, quoting Vladimir Putin himself. The phrase describes Russian forces without insignia and country affiliation signs that became known as 'little green men' after Putin publicly joked about unidentified soldiers in Crimea ('Putin Says He Told Officials Of Decision To Take Crimea In February 2014', 2015).

In 2014, the fighting in the Donetsk region between the separatists and the Ukrainian army led to multiple casualties and destruction. In April, Vladimir Nikitin, a Russian Duma member, summed up the prevailing view in Russia on the possible future of Eastern Ukraine: 'There are two options out of this crisis: either a federalization of Ukraine or the annexing of Eastern Ukraine to Russia' ('Ukraina: Vostok Vzbuntovalsya', 2014). However, Russia has repeatedly denied accusations by Kiev and the Western governments about equipping and supporting the rebels in Donetsk and Luhansk, as well as the presence of Russian troops on the Ukrainian territory (Smith-Spark & Pleitgen, 2015).

At the same time, the international community was trying to mediate the warring sides. On April 15, 2014, the Russian media reported that American troops would be placed in Eastern Europe ('Kiev atakuet yugo-vostok i 'sushit' Krym', 2014). On April 17, 2014, the rebels and the Ukrainian government managed to negotiate and sign an agreement in Geneva that stipulated that the separatists in Slavyansk and Luhansk must stop fighting, in return receiving guarantees that they would not be attacked by the Ukrainian military. The Russian media reported at that time that the agreement might not be an efficient measure, pointing to possible mercenaries from outside of Ukraine being utilized to fight against the separatists ('Mira – net', 2014). The cease-fire agreement was badly needed, because more than one million people had been displaced from

Eastern Ukraine, according to the United Nations Office of the High Commissioner for Human Rights (Smith-Spark & Pleitgen, 2015). By the end of the year, the military actions slowed down during Christmas and the New Year's holidays, and even Kiev was unusually 'dark and quiet' on New Year's Eve, lacking the usual Christmas lights, fireworks, and celebrations (Jarabik, 2015).

One of the greatest tragedies of this conflict was a Malaysian aircraft (Malaysia Airlines MH17) that was shot from the ground on July 17, 2014, killing all 298 people on board. The aircraft, which was flying from Amsterdam (the Netherlands) to Kuala Lumpur (Malaysia), fell near the city of Donetsk. The Western commentators and the media reported that it was Russia that armed and equipped Eastern Ukraine's soldiers, including 'the missile system that shot down the Malaysian airliner'. The Boeing was shot down by a missile that the Ukrainian government identified as a Russian 'Buk' anti-aircraft system, allegedly supplied by the Russian government to the Eastern Ukrainian separatists. Ukraine's Prosecutor-General Viktor Shokin initiated an investigation that was supposed to be finished by the end of 2014, but so far no definitive conclusion has been announced ('Ukrainian Chief Prosecutor Say Separatists Shot Down Malaysian Airliner', 2015). The prosecutor was quoted as saying: 'But the weapon came from Russia and returned to Russia'. He also lamented that investigators had poor access to the crash site, because the site was controlled by separatists ('We Have No Buk Systems That Could Have Been Used for Downing Malaysian Plane – Militia', 2015).

These accusations were denied both by the Eastern Ukrainian separatist government (of the self-proclaimed Donetsk People's Republic) and the Russian government:

> A Buk is a system composed of four to eight vehicles. To use such a system to shoot down a commercial airliner flying at an altitude of 10,000 meters one has to be a high-class specialist on such systems. We don't have such equipment or such specialists.
>
> ('We Have No Buk Systems That Could Have Been Used for Downing Malaysian Plane – Militia', 2015)

On the contrary, they blamed the Kiev forces, accusing them of shooting the plane with a heat-seeking missile on a fighter plane or a Buk ('We Have No Buk Systems That Could Have Been Used for Downing Malaysian Plane – Militia', 2015). So far, neither side has taken the blame for the accident. In October 2014, *Der Spiegel* published a report by European security analysts that concluded that the missile was fired by the rebels. Representatives from Donetsk and Luhansk again denied the accusations (CSIS, 2015).

The Malaysian aircraft was not the only plane shot in the conflict, although other planes were military, not civilian. A month before the Malaysian aircraft was gunned down, separatists shot a Ukrainian military jet Il-76, on June 14, 2014, killing all 49 people on board. The Ukrainian government categorized and investigated it as a terrorist act. The self-declared government of the People's

Republic of Luhansk admitted that they were responsible for this incident. In support of Kiev, the American government issued the following statement:

> [We] condemn the shooting down of the Ukrainian military plane and continue to be deeply concerned about the situation in eastern Ukraine, including the fact that militant and separatist groups have received heavy weapons from Russia, including tanks, which is a significant escalation.
>
> (A. E. Kramer, 2014)

In Ukraine, Sergei Lyovochkin, a member of Ukraine's Verhovna Rada (parliament) and the former chief of staff to President Viktor Yanukovych alluded to the fact that Ukraine remained in limbo, and the concept of 'One Ukraine' seemed unattainable. He continued that other nations were on the side of Ukraine, citing American aid, France's refusal to sell the Mistral aircraft carrier to Russia, and even Kazakhstan's offer to mediate peace negotiations between Ukraine and Russia (Lyovochkin, 2015).

Because of Western disapproval, Russia started to feel the isolation from the leading nations, exacerbated by Western sanctions. At the G-20 Summit in November 2014, state leaders explicitly expressed their condemnation of the conflict: Barak Obama called Russia's involvement in Ukraine a 'threat to the world', and Canada's Stephen Harper agreed to shake Vladimir Putin's hand only if Putin 'gets out of Ukraine' (CSIS, 2015). In the same month, France postponed the Mistral ship delivery to Russia, citing Francois Hollande who said that the 'current situation in eastern Ukraine still does not permit the delivery of the first Mistral-class amphibious assault ship' (CSIS, 2015). The whole context of Russia–West hostility significantly reduced the chance of the Eastern Ukrainian conflict to be resolved peacefully.

Western analysts argued that the Russian government provided the separatists not only with weapons but also with help by Chechen mercenaries (such as the Dzhokhar Dudayev international peacekeeping battalion) and the regular Russian army, with no regalia, posing as 'little green men' ('Senate Foreign Relations Committee Hearing', 2015). However, in his public statements, Ramzan Kadyrov, head of Chechnya, urged to put an 'end to fighting in the south-east of Ukraine' ('Law Enforcement; Kadyrov: Murder of Chechen Man Fighting Alongside Ukraine Army Ordered by SBU and CIA', 2015). Petro Poroshenko admitted that by June 2014, 65 per cent of the Ukrainian armor was destroyed. He mobilized the population, summoning 250,000 men (Mercouris, 2015). Both conflicting parties kept their heavy weaponry, despite the previous commitment to withdraw it (Barker & Cullison, 2015).

On September 5, 2014, the parties signed the Minsk Protocol, which was supposed to end the violence and 'committed the government to extensive decentralization of the Donbass', in order to achieve a peace settlement and hold elections (Mercouris, 2015). On the rebels' side, Donbass leaders Zakharchenko and Plotnisky were the representatives of the people of the Donbass. In November 2014, these leaders were elected in the Donbass elections, while the elections were not recognized as fair by the Ukrainian government (Mercouris, 2015).

The use of force restarted in Donetsk at the beginning of 2015, with no end in sight. Donetsk airport was bombed into ruins. On January 22, 2015, eight civilians were killed at a bus stop in Donetsk, as well as ten Ukrainian soldiers. The separatists and the Ukrainian government blamed each other for what Ukraine's Foreign Minister Pavlo Klimkin called a 'terrorist attack'. In response, the Russian government called it a 'crude provocation aimed at sabotaging efforts to reach a peaceful solution to the Ukrainian crisis' (Shchetko, 2015). Less than a week later, on January 27, the Ukrainian military reported casualties of nine soldiers, while dozens were wounded around the cities of Donetsk and Mariupol. All this violence was happening despite a cease-fire agreed in September 2014 ('Nine Soldiers Killed in 24 Hours as Violence Surges in Ukraine', 2015). Ukraine's national defense council announced that there were 8,500 Russian regular troops in the conflict area, while another 52,000 were stationed on the Russian side of the Russia–Ukraine border. The Russian government denied these figures, accusing the Ukrainian side of manipulation and exaggeration (Vinocur, 2015).

In January 2015, Kremlin spokesman Dmitry Peskov and Russia's Foreign Ministry officials stated that Kiev was 'restarting the fighting'. They warned that a 'planned summit of the Russian, Ukrainian, German, and French leaders to discuss a way to end the violence is thus unlikely' (Barker & Cullison, 2015). Nevertheless, a peace proposal by President Putin, sent to Ukraine in that month, was described by President Poroshenko as a 'sign from Russia that they [Russia] are ready to [put] pressure on these terrorists' (Barker & Cullison, 2015).

On February 12, 2015, the heads of state from France, Germany, Russia, and Ukraine came together again to negotiate the end of the conflict. The negotiations, which Petro Poroshenko described as 'no good news', did not bring many new restrictions or developments, aside from a cease-fire and heavy weaponry withdrawal scheduled for February 15, 2015 ('From Cold War to Hot War', 2015). Such promises had been broken before. The Western media evaluated Vladimir Putin's actions as an attempt to 'bring the collapse and division of Ukraine', while Russia felt threatened by NATO and the EU encirclement ('From Cold War to Hot War', 2015). In the Eastern Ukrainian view, the Donbass separatists were fighting not only the Ukrainian military, but 'against a corrupt Western way of life in order to defend Russia's distinct world view' ('From Cold War to Hot War', 2015).

The economic implications of the Eastern Ukraine conflict

The economic significance of the breakaway Donbass region of Ukraine is very high, as Donetsk and Luhansk produce about 15 per cent of Ukraine's GDP and 25 per cent of the country's industrial output (CSIS, 2015). The region, however, has not been doing too well since the military actions.

Even though the annexation of Crimea and the start of the secession movement in Eastern Ukraine created a sense of euphoria among the pro-Russian population of Ukraine, one year after the 2014 February revolution, the popular sentiments in Donetsk and Luhansk became quite pessimistic, as the people were de-facto abandoned by both Russian and Ukrainian governments, while the fighting continued and attempts to establish a cease-fire had failed. Russian journalists in Donetsk reported about 'apartment buildings bombed into rubble on the city outskirts and long lines all over the city for food, humanitarian aid and medicine'. The separatist government has been struggling to provide basic necessities, some of which came from Russian food aid. Even a Ukrainian oligarch Rinat Akhmetov provided humanitarian aid and distributed food to help people in his native Donestsk (Silchenko, 2015).

The conflict in the Eastern Ukraine (the Donbass region and its cities Donetsk and Luhansk) has led to a humanitarian crisis, as most health care workers had to leave. The 5 million people of the region have been left exposed to untreated infections, such as measles, polio, and TB, because of the lack of vaccinations and appropriate treatment. According to the World Health Organization (WHO), '[i]nsecurity, displacement and cold weather, combined with the poor state of the country's health system, means that basic health care is out of the reach of many people' (Sarchet, 2015). The situation is further complicated by the destruction of the electricity infrastructure and water supply during the military operations (Sarchet, 2015).

The administration of the Donetsk People's Republic, led by Boris Litvinov, has been, in their own words, 'making progress in their state-building' and working on rebuilding the economy and social services. They have even planned to introduce a new currency (Silchenko, 2015). So far, however, the situation in the People's Republic of Donetsk (DNR) has been dire, with thousands of refugees fleeing to Russia (Jarabik, 2015). The Russian government issued a statement of support for the November 2014 elections in Donetsk and Luhansk (CSIS, 2015). In his turn, the 'rebel ideologue in Donetsk', Alexander Borodai, continued to issue statements that the Donetsk separatists would be willing to fight on behalf of Moscow ('From Cold War to Hot War', 2015).

The casualties and destruction during the Eastern Ukrainian conflict were chilling the public attitude in Russia toward the separatists. In total, the United Nations Office for the Coordination of Humanitarian Affairs (OCHA) announced that 1.2 million people were internally displaced in Ukraine, with about 780,000 having fled Ukraine to neighboring countries ('UN Relief Wing Says 1.2 Million Ukrainians Uprooted by Conflict, Calls for More Funding', 2015). According to the United Nations High Commissioner for Refugees (UNHCR), as of April 2015, about 30,000 refugees from Ukraine relocated to Russia. ('UNHCR Says Around 30,000 Ukrainian Refugees are in Russia – Foreign Ministry', 2015). As the refugees were arriving to Russian regions and receiving support from the Russian government, the local population felt that this money could have been used to support the Russian population instead. In January 2015, even Ukrainian President Petro Poroshenko and the Western media noted that Russian public support for the rebels 'appeared to wane in recent months'. As a result, in the Russian official rhetoric, the Eastern Ukraine issue has been separated from the Crimea issue, and the Russian government has expressed no intention to annex Eastern Ukraine, especially in light of the Western sanctions that brought the Russian economy into a recession (Barker & Cullison, 2015).

Western sanctions

After the February 2014 revolution, the United States has been continuously working on strengthening relations with Ukraine:

> The United States attaches great importance to the success of Ukraine's transition to a democratic state with a flourishing market economy… U.S. policy is centered on realizing and strengthening a democratic, prosperous, and secure Ukraine more closely integrated into Europe and Euro-Atlantic structures… It also emphasizes the continued commitment of the United States to support enhanced engagement between the North Atlantic Treaty Organization (NATO) and Ukraine.
>
> ('U.S. Relations with Ukraine', 2015)

These elements of cooperation – 'the areas of defense, security, economics and trade, energy security, democracy, and cultural exchanges' – have been covered in the U.S.–Ukraine Charter on Strategic Partnership and executed by the Strategic Partnership Commission ('U.S. Relations with Ukraine', 2015). The United States and Ukraine have been working on normalizing their trade relations, following the repeal of the Jackson-Vanik amendment for Ukraine, and the implementation of the U.S.–Ukraine bilateral investment treaty, which was promoted by the U.S.–Ukraine Council on Trade and Investment ('U.S. Relations with Ukraine', 2015).

Simultaneously, in support of Ukraine, the Obama administration imposed economic sanctions on Russia. In response to the sanctions, the Russian government announced that the sanctions were an opportunity for Russia to re-orient its economy from reliance on foreign investors to import substitution strategies (Gudkova, 2014).

In January 2015, a commission on Ukraine–NATO cooperation convened to discuss a shelling of Mariupol, allegedly by the Grad missile from the separatist-controlled territory. American Vice President Joe Biden and Ukrainian President Petro Poroshenko publicly condemned the restart of military actions in Eastern Ukraine, and the United States and Ukraine agreed on expanding the sanctions against Russia. American President Barak Obama supported this action in his statements, urging the European Union to cooperate with Ukraine and the United States on this issue, in order for the sanctions to have an effect ('Nine Soldiers Killed in 24 Hours as Violence Surges in Ukraine', 2015). To follow up, U.S. Secretary of State John Kerry visited Kiev in order to meet with Ukrainian President Petro Poroshenko, Prime Minister Arseniy Yatsenyuk, and Foreign Minister Pavlo Klimkin, as well as several Ukrainian parliamentarians. This visit, according to the U.S. Department of State, has shown 'the United States' steadfast support for Ukraine and its people' (Psaki, 2015c).

President Petro Poroshenko has positively evaluated the effect of the sanctions: in January 2015, he said that 'sanctions are working' on Russia, in order to 'keep [Russia] at the negotiating table'. He added, though, that the sanctions did not stop Russia from sending 1,000 additional troops to Ukraine. The Russian government denied his statement, calling it 'absolute nonsense' (Barker & Cullison, 2015).

The Russian government has also denied any negative effect of the sanctions on the Russian economy and refused to change course in dealing with Ukraine. For example, in January 2015, Deputy Foreign Minister Sergey Ryabkov stated that Russia would not negotiate the lifting of sanctions:

> We have repeatedly talked about the illegitimacy of sanctions against Russia. We confirm that under no circumstances would we discuss with the authors of this policy and those who follow it the criteria for the lifting of any sanctions.
>
> ('Russia's Support for Ukraine Will Not Be Endless – Russian PM', 2015)

According to the American government, the government-controlled Russian media have been conducting 'a virulent anti-Western and particularly anti-American campaign for years', and this campaign especially intensified when the Western sanctions were imposed on Russia ('Senate Foreign Relations Committee Hearing', 2015). The Western sanctions on Russia have been accompanied by the West's financial help to Ukraine, the International Monetary Fund provided Ukraine with a $17-billion loan ('From Cold War to Hot War', 2015). The Fund later promised to provide $15 billion more (Barker & Cullison, 2015).

The United States, as the least dependent on Russia in economic and energy security issues, has been a much stauncher proponent of the sanctions against Russia. The European Union, because of the dependencies explained in Chapter 4, has been less united on the extent and the issue-areas of the sanctions. Some European leaders have been concerned that if the Russian economy collapses because of the sanctions, it might have severe repercussions for the European economic system itself. They compared this possibility to the global financial crisis of 1997–1998 that started with a collapse of a single currency in one country – Thailand – and the contagion brought down the economies around the world. What has been happening in Russia in 2014–2015 is just as bad as the situation in Thailand in 1997: as oil prices fell in half, the ruble lost half of its value, aggravated by the effect of Western sanctions ('Decline in Russia's GDP to Cause Domino Effect in the Region – EBRD', 2015).

At a top-level meeting in December 2014, European leaders expressed concerns about their economies that might be dragged down by the deteriorating Russian economic situation. French President François Hollande even said that the sanctions were not productive at all and that they must stop. Russian President Vladimir Putin has repeatedly reinforced this point himself, emphasizing global economic interconnectedness and the World Trade Organization's rules of free trade (Marshall, 2015).

Even the United States did not shut down all areas of U.S.–Russia cooperation: some aspects remained unaffected by the Western sanctions. One such area is the International Space Station, in which cooperation with Russia has been vital. In January 2015, an American contractor for NASA, Orbital Science, purchased $1 billion worth of rocket engines from Energia – Russia's leading space equipment producer – and the deliveries were supposed to start in half a year after the signing of the contract ('1 Billion Dollar Deal: US Firm to Buy 60 Rocket Engines from Russia's Energia', 2015).

Russia's imperfect market structure has alleviated the effect of Western sanctions on Russia's elite circle: in October 2014, the Russian Duma passed the law, commonly described by the media as the 'Rotenberg law', which gave 'government compensation for Russians whose property is seized in a foreign jurisdiction as a result of sanctions' (CSIS, 2015). Simply speaking, the political and business elite that has been affected by the sanctions will be compensated from the Russian federal budget. The sanctions, in this case, became toothless, as they did not affect the decision-makers and only hurt the Russian general public.

In response to Western sanctions, Russia imposed its own import ban (sanctions) on Western countries in several product groups. On August 7, 2014, Russia imposed an import ban in the agricultural sector on the following products from the United States, Canada, Australia, and the European Union: various kinds of meat, chicken, fish, milk, vegetables, fruit, and some processed foods. According to the Russia Agribusiness Report, the main 'losers' of the Russian sanctions will be exporters of American chicken and Norwegian fish ('Russia Agribusiness Report – Q3 2015', 2015).

In 2015, one year after the Western sanctions were imposed on Russia, western analysts have reported that the Russian economy bounced back and the ruble started to appreciate. Russia's GDP is predicted to grow at the rate of about 3.5 per cent per year, regardless of the level of oil prices (Powell, 2015). One of the reasons for this growth has been attributed to the import substitution strategy: i.e., the reliance on the domestic production of goods and services. Another reason is that the cheaper ruble makes Russian exports cheaper for those products that have not been banned by Western sanctions, such as commodities. At the same time, the export revenues for commodities (oil and natural gas in particular) come to the country in dollars, and the strong dollar substantially increases export revenues converted into rubles (Powell, 2015).

Historically, however, the import substitution strategy has proved itself to be less efficient in the long-term than export-oriented strategies. A whole branch of the political economy literature has studied these strategies in much detail (Gary Gereffi, Stephan Haggard, Barbara Stallings, and many others). The so-called Import Substitution Industrialization (ISI) prevailed in Latin America in the 1980s and the 1990s: the economic policy in those countries was based on the ideas of a 'strong state and a relatively closed economy'. In East Asia, on the contrary, the economy was extremely export-oriented (Stallings & Peres,

2011). As a result, the Asian countries achieved a better level of economic growth and development.

One might certainly argue that Russia right now has fewer export opportunities than the East Asian countries did, because the Asian 'tigers' did not have economic sanctions imposed on them. However, the historic experience of the ISI and the export-oriented development is still worth studying, as the conditions in Russia are currently somewhat similar to those in Latin America in the 1980s (a strong state, weak currency, and dependency on imports).

The Western economic sanctions have been accompanied by a severe deterioration of Russia–NATO relations. In April 2014, Russia accused the military alliance of strengthening NATO troops in Eastern Europe, despite the existing agreements (such as the 2002 Roma Declaration) not to do so. As a result, Russia even withdrew its NATO envoy. NATO Secretary-General Anders Fogh Rasmussen denied the plans of expanding NATO's military presence in Eastern Europe in response to the Ukraine conflict. He instead referred to the Russian government's rhetoric as 'propaganda and disinformation' ('Russia Rejects Kyiv's Sniper Claims; NATO Dismisses Moscow's Accusations', 2014). He continued, however, that despite the current disagreements with Russia, NATO will still be able to use Russian air space (a part of the so-called 'northern supply line' for the American troops) for the U.S. military transit to Afghanistan ('Russia Rejects Kyiv's Sniper Claims; NATO Dismisses Moscow's Accusations', 2014).

Russian oil and gas transit to Europe

The main concern from the events above is how the situation in Ukraine has changed the European energy policies and Russia's energy trade with the West.

Ukraine and Russia are dependent on each other in the energy security domain. Russia needs Ukraine for oil and gas transit to Western Europe, and Ukraine needs Russia for transit fee revenues and natural gas supply. Despite this mutual dependence on natural gas trade and transportation, Ukraine and Russia have had multiple disputes about natural gas prices and the conditions of transit. Russia has repeatedly attempted to bring the price for the gas intended for Ukraine up to the general European level. After the February revolution of 2014, in April of the same year, Russia's energy giant Gazprom again tried to raise the price to $485 per 1,000 cubic meters. Ukraine's negotiation position was weak, as the country owed $2 billion to Russia for the gas deliveries. As in the previous instances, the Ukrainian government described Russia's attempts to charge a common market price as a 'political' advantage and an energy weapon to weaken Ukraine's economy ('Russia Rejects Kyiv's Sniper Claims; NATO Dismisses Moscow's Accusations', 2014).

An area of contention between Russia and the European Union has been the modernization of Ukraine's natural gas transportation system, actively assisted by the EU. The Russian government has been concerned that the EU was planning to isolate Russia, not allowing it to participate in this project. The European partners assured the Russian government that it was not the case. In winter of 2010, Russia offered to modernize the Ukrainian distribution system jointly with the EU, in order to ensure a stable supply of 19.5 billion cubic meters of natural gas to Europe via Ukraine. As Ukraine had been known to use the Western Europe-bound natural gas for its own domestic needs, Russia was concerned that the amount of natural gas that would actually reach Western Europe could fall short. Ukraine was waiting for a decrease in natural gas prices and had not purchased enough gas reserves to cover its domestic needs. That situation was leading to another gas crisis that could only be averted if the EU and Russia monitored the Ukrainian gas reserves jointly (Zygar & Grib, 2009).

Russia's attempts to increase natural gas prices for Ukraine is not surprising, though, if one follows the domestic price dynamics in Russia. Urged by the WTO, the EU, and global economic liberalization tendencies, Russia has increased domestic gas prices several times, promising to gradually bring them to the level of export prices (minus export duties and transit fees). This will be a challenging task for the Russian government, considering that Russian households might be unable to pay as much as the Europeans who have substantially higher incomes. In addition to this, if Russian domestic prices become equal to the price of natural gas in Europe, Gazprom will be less motivated to export, since domestic profits will be equal to potential export profits (Papava et al., 2009, p. 42). Therefore, the 'one-size-fits-all' liberalization of Russian natural gas prices might not be favorable for European energy security.

Ukraine's dependency on Russian natural gas is simply too high for Ukraine to ever feel not pressured into concessions and price disputes. Ukrainian President Victor Yanukovych said in September 2013 that the 'main problem in Ukraine's relations with Russia had always been Russia's use of energy to pressure Ukraine' ('Economic Policy; Yanukovych: Dialog Between Ukraine, Russia, EU on Trade Issues Possible in Near Future', 2013). A year later, in October 2014, the pro-European Ukrainian Prime Minister Arseniy Yatsenyuk was still negotiating with Russia over natural gas deliveries. Russia was demanding prepayments for natural gas, while Ukrainian Energy Minister Yuriy Prodan complained that Russia's demands breached the previous terms of Russia–Ukraine contracts. The negotiations were complicated by the Ukrainian debt of $2 billion to Russia for previous shipments of natural gas, reducing Ukraine's bargaining power in all subsequent negotiations. By the end of the negotiations, the two countries signed a contract, agreeing on the price of $378 per 1,000 cubic meters, with the European Commission as a guarantor for the Ukrainian side (CSIS, 2015).

For Ukraine, Russia's requirement of prepayments was not a surprise: Russia had tried to switch to the prepaid system for a while. On one of such occasions, in April 2014, Vladimir Putin in his televised speech explained how Russia planned to build further relations with Ukraine in the natural gas trade. He stated that Ukraine had not been paying for Russian natural gas for a long time, and therefore Russia planned to switch to prepaid deliveries ('Kiev atakuet yugo-vostok i "sushit" Krym', 2014). As negotiations in October 2014 were not bringing desired results, Vladimir Putin announced that the transit risks related to Ukraine would make Russia reduce the natural gas volume transported to Europe via Ukraine, because of Ukraine's 'illegal taking' of natural gas intended for Europe (CSIS, 2015). In response, Ukraine signed a contract to import natural gas from Norway's Statoil (the price and the volume of the contract were not made public) (CSIS, 2015).

Understanding that Ukraine might eventually switch to other suppliers, the Russian government has also been planning to switch to other pipelines, such as Nord Stream. During the conflict, Russian gas deliveries for Ukraine were shut off for almost half a year and were resumed only in December 2014, after the two governments agreed on the price and Ukraine found money to prepay for 1 billion cubic meters. At the same time, Russia agreed to continue supplying coal and electricity to Ukraine (CSIS, 2015).

Russian Prime Minister Dmitry Medvedev said, however, that 'Russia's support for Ukraine will not be endless'. He continued to insist that Ukraine must pay $2.44 billion for the gas that had already been delivered. He announced that Ukraine would have to start paying the full European price for natural gas starting from April 2, 2015, when the special discounted price of $365 for 1,000 cubic meters expired ('Russia's Support for Ukraine Will Not Be Endless – Russian PM', 2015).

For Ukraine, to pay the $2 billion debt to Russia has been close to impossible, as the economy has been quickly going down since the start of the conflict with Russia. Ukraine's foreign currency reserves fell sharply to $7.5 billion, with an inflation rate of almost 25 per cent and a 2014 budget deficit of 8.2 per cent of GDP. That left little money for Ukraine's own survival, let alone for paying off its debt for Russian oil and gas (Jarabik, 2015). In October 2014, the World Bank estimated that the Ukrainian economy might start growing only in 2016, while contracting by '8 per cent in 2014 and 1 per cent in 2015' (CSIS, 2015).

Ukraine asked for economic assistance, and the West sent billions of dollars in financial aid. The International Monetary Fund (IMF) provided a $15 billion loan. Individual governments sent their own help: the United States provided $2 billion, Germany provided half a billion dollars, and the European Union provided $2 billion. However, it remains to be seen if this help will actually save the plunging Ukrainian economy (Jarabik, 2015). Ukrainian Prime Minister Arseniy Yatsenyuk said in 2015 that the country needs at least $30 billion, for the Ukrainian economy not to collapse (CSIS, 2015).

One could expect that the plunging oil prices and the ruble's crumbling value in 2014 would diminish Russia's bargaining power, and the Russian government would make more concessions to Ukraine. However, the economic recession in Russia, caused by low oil prices, did not change Vladimir Putin's negotiation style with Petro Poroshenko. In his interview in January 2015, Poroshenko said that the fall in oil prices and the ruble's value in recent months had not changed President Putin's 'tone in the calls' (Barker & Cullison, 2015).

For Russia, the crisis in Ukraine has created a huge incentive to speed up the pipeline diversification. In addition to Nord Stream (explored in detail in

Chapter 4), another 'sister' pipeline – South Stream – was supposed to be built via Bulgaria. Not surprisingly, however, Western governments objected to the South Stream project and accused the Bulgarian government in its desire to amend EU rules to favor Russia. Because of the controversy, Bulgaria preferred to distance itself from the project, and Vladimir Putin pulled the plug on it ('From Cold War to Hot War', 2015).

After Bulgaria, another surprising 'ally' of Russia in Europe has been European 'green' parties that have 'opposed... shale-gas fracking and nuclear power'. Western commentators claimed that Russia also opposed those two energy sources in Europe, because they would increase European energy independence. As a result, the Western media speculated that the Bulgarian government's refusal to grant Chevron (USA) the rights to use hydraulic fracturing in Bulgaria was because of Russia's meddling ('From Cold War to Hot War', 2015). Such connections between the events seem far-fetched, though, as fracking has been a debated issue in numerous countries, and especially in Europe. The hydraulic fracturing's environmental costs, such as millions of gallons of clean water used per well; toxic chemicals used for fracturing; and the wastewater potentially contaminating aquifers and surface waters, have been important to consider on their own.

What Russia's Gazprom did was urge the European Union to avoid Ukraine as a transit state and to use Turkey instead, diverting natural gas that was previously going via Ukraine. The Russian government proposed a new pipeline, dubbed the 'Turkish Stream', replacing the previously cancelled 'South Stream'. Gazprom CEO Alexei Miller announced that the Turkish Stream, if built, could deliver 63 billion cubic meters of natural gas (Lossan, 2015). The Turkish Stream pipeline is planned to have four lines: one will deliver natural gas for Turkey's domestic needs, and the others will transport natural gas to the European Union. The pipeline will be connected to the Trans-Adriatic Pipeline (TAP) and provide Russia with access to South Europe. TAP, in its turn, will join the Trans-Anatolian Pipeline (TANAP), providing natural gas to such countries as Turkey, Italy, Albania, and Greece ('Emerging Europe Oil and Gas Insight – May 2015', 2015, pp. 1–2). Experts expressed concerns, though, whether TAP has enough capacity, on which Gazprom is now relying in its calculations for the future of the Turkish Stream. If TAP is limited to two lines, Gazprom might not have enough capacity for its natural gas transportation ('Emerging Europe Oil and Gas Insight – May 2015', 2015, pp. 1–2).

Russia's plans for new pipelines such as the Turkish Stream have been happening amidst growing European attempts to reduce the dependency on Russian gas. The Russian media described such European policies as a 'redrawing' of Europe's energy map, in order to push Gazprom from the European market. The Russians quoted European policy makers who were planning to switch to imported American shale gas, which had been promised by U.S. President Barak Obama. The Russian media emphasized that because of future LNG supplies to Europe, the United States would be implementing European dependency on American gas. Referring also to the American support of Ukraine, the Russian media focused on what they perceived as Obama's attempts to 'mute' the failure of the American policies in Afghanistan, Iraq, Iran, Libya, and Syria, and to increase the American influence in Europe (Kostikov, 2014).

The Russian media also criticized Europe for not trying to understand Russia's position, as the country made a leap from the Soviet Union to a capitalist system in a little over 20 years. Another area of criticism was NATO bases around Russia that 'made Vladimir Putin draw a geopolitical "red line" in Ukraine'. Russia denied accusations of planning 'aggression' against Europe and called it the European 'fantasy' (Kostikov, 2014). The Western governments, however, have been very cautious of Vladimir Putin's policies:

> He has built gas pipelines to Western Europe around Ukraine and even ally Belarus so that he can use gas as a weapon against these countries, while maintaining access to his wealthy customers in the West. He has hired shameless senior European officials to work as front men in his companies.
> ('Senate Foreign Relations Committee Hearing', 2015)

As Russia's relations with the West have been at their lowest point since the end of the Cold War, the Russian government has sought to diversify its customer base for natural gas and oil exports. As a result, a $400 billion natural gas deal was signed in May 2014 with China. Although Mikhail Margelov, head of the Federation Council's international affairs committee, announced that there was 'no connection between the deal and the deterioration in relations between Moscow and Brussels', such a connection has been established by Western commentators ('Reaction Mixed to Russia's 400bn-Dollar Gas Deal with China', 2014).

Under the deal, Russia will start natural gas deliveries to China in 2018, at the estimated price of $350 to $390 per 1,000 cubic meters (the actual price remained confidential) ('Reaction Mixed to Russia's 400bn-Dollar Gas Deal with China', 2014). The deal was signed on the assumption that the necessary infrastructure (new pipelines) would be built: one such pipeline – the Power of Siberia – has a planned capacity of 38 billion cubic meters and will go from West Siberia to Yakutia and Irkutsk, with the final destination in Khabarosk ('Emerging Europe Oil and Gas Insight – May 2015', 2015, p. 3).

Russia has been expanding its cooperation with China even beyond trade deals. The Russian government has already discussed possibilities for Chinese companies to acquire major stakes in Russian oil and gas companies: something that has not been allowed to American companies for more than a decade. Earlier in 2015, Deputy Prime Minister Arkadiy Dvorkovich stated that there were 'no political barriers' on the Russian side to offering 'as much as a 49–50-per cent stake' in some companies and projects to Chinese companies, such as China National Petroleum Corporation (CNPC) ('Emerging Europe Oil and Gas Insight – May 2015', 2015, p. 4).

Domestic interest groups in Russia

Not everyone in Russia has agreed with the Russian government on the Russia–China deal, Ukraine, and Crimea. A vocal opposition movement has published multiple reports and staged protests against the mainstream policy line by the Putin administration.

On the China deal, one of the most influential and vocal opposition leaders in Russia was Boris Nemtsov, assassinated on February 27, 2015, in Moscow (as of September 2015, his murderers have not been found). In his capacity as First Vice-Prime Minister during the Yeltsin administration in the 1990s, Boris Nemtsov worked on deals with China, such as nuclear plant building by Russian companies in China and selling Russian planes to Chinese companies, among other areas (Kuznetsova, 1998). Historically, Nemtsov was not an opponent of all cooperation with China.

However, he was extremely critical of the 2014 Russia–China gas deal. Publishing his analysis mostly on social media, such as Facebook, he stated that the cost of infrastructure and delivery for the project would be about $155 per 1,000 cubic meters of natural gas. In addition to that, Gazprom would pay about $140 of taxes per 1,000 cubic meters, making:

> a profit of 55 dollars at a sale price of 350 dollars or 16 per cent of revenue. The contract revenue is 400bn [dollars]. Gross profit 64bn. Cost of serving loans 77.5bn. Overall loss to Gazprom from the deal is 13.5bn. Such is the contract of the century. Of course, Gazprom won't work at a loss. It will demand tax breaks, of about 15-20bn dollars.
>
> ('Reaction Mixed to Russia's 400bn-Dollar Gas Deal with China', 2014)

In short, Nemtsov calculated that Russia might be trading with China at a loss, and the economic feasibility of the contract would improve only with substantial tax breaks for oil and gas companies.

Boris Nemtsov was not the only critic of the deal. Another influential opposition figure, Aleksey Navalny, has agreed with Nemtsov that the deal might not be profitable for the Russian budget, because the initial construction cost estimate of $55 billion will most likely be exceeded multifold. He compared the Russia–China deal with the ESPO pipeline: ESPO cost $16 billion to build, out of which $4 billion 'was directly stolen', according to Navalny ('Reaction Mixed to Russia's 400bn-Dollar Gas Deal with China', 2014). The Russian government has denied these accusations. Navalny has demanded that the Russia–China deal be made public, so that the voters can monitor the expenses and revenues.

Both Nemtsov and Navalny were very critical of the Russian government's policy in Crimea and Ukraine. Another opposition group has been dissatisfied with the reunification of Crimea into the Russian territory – the Crimea Tatars ('Putin Says He Told Officials Of Decision To Take Crimea In February 2014', 2015).

This is not the first time in history that the Tatars opposed the status quo in Crimea. Back in 1992–1995, shortly after the dissolution of the USSR, Crimea already tried to secede from Ukraine. The Russian-speaking population organized a referendum that confirmed their right to hold Russian citizenship. They even established a government and passed a constitution, independent of Ukraine (Rupert, 1995). However, that secession movement was peacefully 'defused' by the Ukrainian government and Ukrainian President Leonid Kuchma via negotiations and compromises, as well as administrative steps such as abolishing the newly adopted Crimean constitution and stripping of power of the newly proclaimed President of Crimea (Rupert, 1995).

In those negotiations between the pro-Russian separatist government and Kiev, the Crimean Tatars (ten per cent of the Crimean population) took the Ukrainian government's side, simultaneously making their own demands for more political and economic power, their own cultural institutions and educational opportunities. For example, they wanted quotas in the Crimean parliament and various government agencies, as well as full voting rights that could come only with Ukrainian citizenship, which had been denied to many of them because of the exile during the Soviet time (Rupert, 1995). Although Tatars had lived on the peninsula for five centuries, they had to compete with the Russian population for jobs and resources in the poor and unstable post-Soviet Crimea.

Crimea Tatars have had historical grievances against Russia as a successor to the Soviet Union, because in 1944, the Soviet leader Joseph Stalin forcefully moved the whole Tatar population to Central Asia, on the spurious charges of Tatar collaboration with the Nazis. It was part of Stalin's policy of moving small ethnic communities (Tatars, Greeks, Bulgarians, and Armenians) from Crimea (M. Kramer, 2014). The Tatars stayed in Central Asia for almost half a century after the relocation, and they were able to return to Crimea only at the end of the 1980s, during Mikhail Govbachev's perestroika and liberalization (Rupert, 1995).

In 2014, Tatars were already 20 per cent of the Crimean population, and they made demands to the Russian government similar to those they had made to the Ukrainian government in 1995. This time, they again felt that their rights were not protected in Crimea. In the spring of 2015, the media around the world reported about the Tatar TV station, ATR, being closed by the Russian government, and the station's application for re-registration had been repeatedly denied. ATR was the 'last independent Tatar station in Crimea'. Shortly after that the Crimean authorities closed another Tatar media outlet – the Meydan FM radio station. The Tatars have repeatedly complained that their schools and religious institutions are regularly visited by the authorities, with the intent to intimidate them (MacFarquhar, 2015). Crimean Prime Minister Sergei Aksyonov denied the accusations in human rights violations and limiting the Tatars' freedoms. He said that the stations had made mistakes in their registration applications. In his turn, he accused the ATR station and Meydan FM in creating the issue intentionally, in order to spread the discord in Crimea (MacFarquhar, 2015). Other Crimean politicians even compared the Tatars with the 'fifth column' – a term for those who were allegedly instructed by the West and the Ukrainian government. The Crimean government media posted pictures of Tatar leaders with Ukrainian President Petro Poroshenko, as evidence (MacFarquhar, 2015).

Conclusion

Russia's conflict with Ukraine has direct implications for European energy security. Ukraine has been diversifying its natural gas supply, signing deals with Norway. The European Union has also been intensifying policies to reduce dependency on Russian natural gas. Russia, in its turn, has turned to China for future energy deals, even though the Russia–China deal of 2014 has been criticized by the Russian opposition movement as not profitable for the Russian budget.

Aside from energy security implications, the military conflict in Eastern Ukraine has been a humanitarian tragedy. About 6,000 people have been killed, and millions have become displaced, both internally and abroad. Tens of thousands of refugees fled to Russia, creating tensions on the Russian territory for financial help and jobs, as Russia has been sliding into a recession because of low oil prices and Western sanctions.

The relationships between the West and Russia, following Crimea's annexation, have been at the lowest point since the breakup of the Soviet Union. A way out of this conundrum could be the West's diversification of energy resources, including renewable energy, so that European countries are not dependent on Russian natural gas as much and the relations might be able to be normalized.

References

1 Billion Dollar Deal: US Firm to Buy 60 Rocket Engines from Russia's Energia. (2015, January 20). *Customs Today*. Retrieved from http://customstoday.com.pk

Barker, T., & Cullison, A. (2015, January 20). Ukraine President Petro Poroshenko: 'Sanctions Are Working' on Russia. *Wall Street Journal.*

Bilozerska, O. (2015). The Day Euromaidan Turned Violent, One Year After. *Euromaidan Press*. Retrieved from http://euromaidanpress.com

Biografia Ideologa Ukrainskogo Natsionalizma Stepana Bandery. (2015). Retrieved from http://ria.ru/spravka/20150102/1041186412.html

Brainin, K. (2014, October 19). Na Ukraine, Otchayanno Soprativlyavsheisya Fascismu, Segodnya Privetstvuyut Nazistov, Marshiruyuschih Po Ulitsam. Retrieved April 25, 2015, from http://www.1tv.ru/news/print/270048

Confectionery; Russia Bans Imports from Ukraine's Roshen. (2013). *Interfax: Russia & CIS Food & Agriculture Weekly*.

CSIS. (2015). The Ukrainian Crisis Timeline: Center for Strategic and International Studies.

Decline in Russia's GDP to Cause Domino Effect in the Region – EBRD. (2015, January 19). *Russia Today*. Retrieved from www.rt.com

Dibrova, V. (2013, April 14). Symposium at Harvard focuses on language politics in Ukraine. *Ukrainian Weekly*, pp. 1–14.

Direct Line with Vladimir Putin on April 17, 2014. (2014). Retrieved April 14, 2015, from http://en.kremlin.ru/events/president/news/20796

Dubrovskaya, L. (2015, March 17). Peskov – Psaki: Net Nikakoi Okkupatsii Kryma, Tema Ne Podlezhit Obsuzhdeniyu. *MK*. Retrieved from http://www.mk.ru/politics/2015/03/17/rf-perebrasyvaet-raketonoscy-v-krym-peskov-kommentiruet-okkupaciyu-poluostrova.html

Economic Policy; Yanukovych Tells Grybauskaite Kyiv Cannot Sign Agreement With EU Because of Russian Pressure. (2013, November 22). *Interfax: Ukraine Business Daily*.

Economic Policy; Yanukovych: Dialog Between Ukraine, Russia, EU on Trade Issues Possible in Near Future. (2013, September 25). *Interfax: Ukraine Business Daily*.

Emerging Europe Oil and Gas Insight – May 2015. (2015) (pp. 1–9). London: Business Monitor International.

English, R. (2013, March 13). Ukraine's Threat from Within. *The Los Angeles Times*. Retrieved from http://articles.latimes.com/2014/mar/13/opinion/la-oe-english-ukraine-neofascists-20140313

Erlanger, S. (1995, June 10). Russia and Ukraine Settle Dispute Over Black Sea Fleet. *New York Times*.

Far-right Ukrainians Mark Anniversary of Nationalist Hero Stepan Bandera. (2015, January 1). *Euronews.Com*. Retrieved from http://www.euronews.com/2015/01/01/far-right-ukrainians-mark-anniversary-of-nationalist-hero-stepan-bandera/

From Cold War to Hot War. (2015, February 14). *Economist*, pp. 19–20.

Goble, P. (2015, January 20). Igor Eidman: Putin Became a Fascist-style Leader in Mid-2014. *Euromadan Press*. Retrieved from http://euromaidanpress.com

Gudkova, V. (2014). Pakhanu – po rukam! V SShA Zreet Novyi Finansovy Puzyr. *Argumenti i Fakti, #17*, p. 19.

Harris, R. (2014, March 17). Putin's Games End Under a Crimean Cloud. *The Charleston Gazette*.

Interfax. (2015, January 20). Medvedev Calls Agreements on Electricity Shipments to Ukraine and Crimea Mutually Beneficial.

Jarabik, B. (2015, January 20). Ukraine Fatigue: To Be or Not to Be (Bailed Out). *Carnegie Moscow Center*.

Kenarov, D. (2015, January 19). Dashed Hopes in Gay Ukraine. *Foreign Policy*.

Kiev atakuet yugo-vostok i 'sushit' Krym. (2014). *Argumenti i Fakti, #17*, p. 10.

Kostikov, V. (2014). Khor Slepyh. *Argumenti i Fakti, #15*, p. 5.

Kramer, A. E. (2014, June 15). Separatists Down Military Transport Jet, Killing 49 in Eastern Ukraine. *New York Times*.

Kramer, M. (2014). Why Did Russia Give Away Crimea Sixty Years Ago? *Wilson Center*. Retrieved from http://www.wilsoncenter.org/publication/why-did-russia-give-away-crimea-sixty-years-ago

Kuznetsova, L. (1998, February 16). Russia and China Resolve Disagreements on Energy Contracts. *ITAR – TASS News Wire*, p. 1.

Language Law Splits Ukrainian Society – OSCE High Commissioner. (2012, July 26). *Interfax: Russia & CIS General Newswire*.

Law Enforcement; Kadyrov: Murder of Chechen Man Fighting Alongside Ukraine Army Ordered by SBU and CIA. (2015, February 6). *Interfax: Russia & CIS Military Information Weekly*.

Leonard, P., & Isachenkov, V. (2014, May 13). Insurgents in Eastern Ukraine Declare Independence. *The Epoch Times*, pp. A10–A11.

Lossan, A. (2015, January 19). Gazprom Throws Down the Gauntlet to Europe over Gas Supplies via Turkey. *Russian Beyond the Headlines*. Retrieved from www.rbth.ru

Lyovochkin, S. (2015). One Ukraine. Retrieved from www.rollcall.com

MacFarquhar, N. (2014, May 10). From Crimea, Putin Trumpets Mother Russia. *New York Times*.

MacFarquhar, N. (2015, April 3). Russia Closes TV Station for Tatars of Crimea. *International New York Times*, p. 3.

Magee, S. K. (2014, March 16). Russia Vetoes U.N. Resolution Calling Crimea's Referendum Invalid. *Kyodo News Service*.

Malysheva, E. (2015, March 15). Glava Minkryma – RBK: 'Sanktsii Sozdayut Massu Problem'. *RBK*. Retrieved from http://top.rbc.ru/economics/15/03/2015/550594419a7947645bd23d8e

Marshall, J. (2015, January 19). Risky Blowback from Russian Sanctions. *Consortium News*. Retrieved January 31, 2015, from www.consortiumnews.com

Mercouris, A. (2015, January 19). Ukraine Goes to War – and Always Will as Long as Maidan Holds Power. *Russia Insider*. Retrieved from http://russia-insider.com

Mira – net. (2014). *Argumenti i Fakti*, *#17*, p. 10.

Mudde, C. (2014, February 28). A New (Order) Ukraine? Assessing the Relevance of Ukraine's Far Right in an EU Perspective. *OpenDemocracy*.

Nine Soldiers Killed in 24 Hours as Violence Surges in Ukraine. (2015, January 27). *The Moscow Times*. Retrieved from http://www.themoscowtimes.com/news/article/nine-soldiers-killed-as-violence-surges-in-ukraine/514943.html

Oberemenko, V. (2014). 'Khaviti Sporit – Pora Deistvovat!'. *Argumenti i Fakti*, *#18*, p. 3.

Onyschuk, B. (2014, March 12). Broken Promises. *National Post*.

Oreshkin, D. (2015, March 17). Rossiya and Krym Cherez God Posle Prisoedineniya. Retrieved from http://snob.ru/profile/8894/blog/89580

Panzeri, A. (2008, August 10). Russia, Georgia Urged to Respect Olympic Truce. *The Gazette*.

Papava, V., Grigoriev, L., Paczynski, W., Tokmazishvili, M., Salikhov, M. Sr., & Bagirov, S. (2009). Energy Trade and Cooperation between the EU and CIS Countries. *SSRN eLibrary*.

Popova, M., & Post, V. (2014, April 09). What is Lustration and is it a Good Idea for Ukraine to Adopt it? *Washington Post*. Retrieved from http://www.washingtonpost.com/blogs/monkey-cage/wp/2014/04/09/what-is-lustration-and-is-it-a-good-idea-for-ukraine-to-adopt-it/

Powell, B. (2015, April 24). What Sanctions? The Russian Economy Is Growing Again. *Newsweek*, p. 164.

Psaki, J. (2015a). *On the Intended Signing of an Agreement Between the De Facto Leaders in Georgia's South Ossetia Region and the Russian Federation*. Washington, DC: Retrieved from http://www.state.gov/r/pa/prs/ps/2015/03/239383.htm

Psaki, J. (2015b). *One Year Later – Russia's Occupation of Crimea*. Washington, DC: Retrieved from http://www.state.gov/r/pa/prs/ps/2015/03/238897.htm

Psaki, J. (2015c). *Press Statement: Secretary Kerry Travels to Kyiv and Munich*. Washington, DC: Retrieved from http://www.state.gov/r/pa/prs/ps/2015/01/237000.htm

Putin Delivers Crimea Annexation Address to Russian Parliament. (2014, March 18). *BBC Monitoring Former Soviet Union*.

Putin Says He Told Officials Of Decision To Take Crimea In February 2014. (2015). Lanham: Federal Information & News Dispatch, Inc.

Putin: Russia Helped Yanukovych to Flee Ukraine. (2014, October 28). *Asia News Monitor*.

Putin: What Was Done in Crimea Was Not in Any Way Different from What Had Been Done in Kosovo. (2014, November 17). *Interfax: Russia & CIS General Newswire*.

Rada Establishes Procedures for Appointing and Dismissing Director of Anti-Corruption Bureau. (2015, February 13). *Interfax: Ukraine General Newswire.*

Reaction Mixed to Russia's 400bn-Dollar Gas Deal with China. (2014, May 22). *BBC Monitoring Former Soviet Union.*

Referendum on Status of Crimea Fits General European Tendencies – Markov. (2014, March 3). *Interfax: Russia & CIS General Newswire.*

Robles, J. (2014, June 6). On D-Day Do Not Forget the Great Patriotic War. *Voice of Russia.* Retrieved from http://sputniknews.com/voiceofrussia/2014_06_06/On-D-Day-do-not-forget-the-Great-Patriotic-War-5478/

Rostovskiy, M. (2015, March 4). 'Sanctions Are A Sign of Obama's Despair?' Don't Make Yourself and the People Laugh. *Moskovskiy Komsomolets.*

Rupert, J. (1995, December 4). Ukraine's Troubles in Crimea Take New Turn; Reconciliation of Russian Separatists Advances, but Now Returning Tatars Make Demands. *The Washington Post.*

Russia Agribusiness Report – Q3 2015. (2015) (pp. 1–100). London: Business Monitor International.

Russia Marks Victory Over Nazi Germany With Military Parade. (2015). Lanham: Federal Information & News Dispatch, Inc.

Russia Rejects Kyiv's Sniper Claims; NATO Dismisses Moscow's Accusations. (2014). Lanham: Federal Information & News Dispatch, Inc.

Russia: Ruble's Fall Sends Cost of World Cup Stadiums Soaring 40 Per Cent. (2015, April 21). *BBC Monitoring Former Soviet Union.*

Russia: Sochi Olympic Games Made USD 53-M Profit – IOC. (2015, March 2). *Asia News Monitor.*

Russia's Support for Ukraine Will Not be Endless – Russian PM. (2015). TASS. Retrieved from http://tass.ru/en/russia/772142

Russian TV News: US Military Training for Ukrainian 'Neo-Nazis'. (2015, April 21). *BBC Monitoring Former Soviet Union.*

S Amerikoi nado derzhat ukho vstro. (2013). *Argumenti i Fakti*, #42, p. 11.

Sarchet, P. (2015, January 20). Ukraine Is Left Struggling as Health Workers Flee. *New Scientist.*

SBU Opens Criminal Case on Russian Bank Financing Terrorists in Eastern Ukraine. (2014, April 15). *Interfax: Ukraine General Newswire.*

Senate Foreign Relations Committee Hearing. (2015). Lanham: Federal Information & News Dispatch, Inc.

Separatist (definition). (2015). *Oxford Dictionaries.* Retrieved from www.oxforddictionaries.com

Shchetko, N. (2015, January 22). Shelling Kills at Least 8 at Bus Stop in Donetsk, Eastern Ukraine. *The Wall Street Journal.* Retrieved from http://www.wsj.com/articles/shelling-kills-13-at-bus-stop-in-donetsk-1421917147

Silchenko, V. (2015, January 20). Residents of War-Torn East Ukraine Left in Deadly Limbo. *Moscow Times.*

Smith-Spark, L., & Pleitgen, F. (2015, February 20). Ukraine Marks Year since Maidan Bloodshed amid Simmering Conflict. *CNN Wire Service.*

Son of Former Ukrainian Leader Viktor Yanukovych Dies in Accident in Russia. (2015, March 22). *The Guardian.* Retrieved from http://www.theguardian.com/world/2015/mar/22/viktor-yanukovych-former-ukrainian-leader-son-dies-accident-russia

Sonne, P. (2014, March 7). Crisis in Ukraine: Crimea Push Gives Putin a Bump in Popularity. *Wall Street Journal.*

Stallings, B., & Peres, W. (2011). Is Economic Reform Dead in Latin America? Rhetoric and Reality since 2000. *Journal of Latin American Studies*, *43*(4), 755–786.

Stepan Bandera Becomes Ukrainian Hero. (2010, January 22). *Interfax: Ukraine General Newswire.*

U.S. Relations with Ukraine. (2015, January 15). Retrieved January 18, 2015, from http://www.state.gov/r/pa/ei/bgn/3211.htm

Uhler, W. C. (2015). The New York Times Sinks to a New Journalistic Low in its Reporting on Ukraine. Retrieved from http://dissidentvoice.org

Ukraina: Vostok Vzbuntovalsya. (2014). *Argumenti i Fakti*, *# 15*, pp. 6–7.

Ukraine Crisis: Crimea's Troubled History is Dictating Modern Day Events. (2014, March 7). *The Telegraph*.

Ukraine Food & Drink Report – Q1 2015. (2015) (pp. 1–142). London: Business Monitor International.

Ukraine Politics: Authorities Target Opposition. (2011). New York, NY: The Economist Intelligence Unit N.A., Incorporated.

Ukraine Protests after Yanukovych EU Deal Rejection. (2013, November 30). *BBC.com*. Retrieved from http://www.bbc.com/news/world-europe-25162563

Ukraine Ratifies the Missile Pact, But Delays Ending Nuclear Status. (1993, November 19). *New York Times.*

Ukraine: Profile: Petro Poroshenko, Ukraine's Presidential Candidate, 'Chocolate King'. (2014, April 11). *Asia News Monitor.*

Ukrainian Chief Prosecutor Say Separatists Shot Down Malaysian Airliner. (2015, March 30). *BBC Monitoring Former Soviet Union.*

UN Relief Wing Says 1.2 Million Ukrainians Uprooted by Conflict, Calls for More Funding. (2015, April 16). *M2 Presswire.*

UNHCR Says Around 30,000 Ukrainian Refugees are in Russia – Foreign Ministry. (2015, April 22). *Interfax: Ukraine General Newswire.*

Vasilyeva, N. (2015, February 16). Russia Still Paying for Sochi Excess; Investors Dump Costly Winter Games Venues onto State; $8.5B Rail Line Going Nowhere. *The Vancouver Sun.*

Vinocur, J. (2015, January 20). Is There No One Who Will Stand Up for Ukraine? *Wall Street Journal.*

We Have No Buk Systems That Could Have Been Used for Downing Malaysian Plane – Militia. (2015, March 31). *Interfax: Russia & CIS General Newswire.*

Weir, F. (2013, February 5). Sochi cha-ching: Putin Defends Most Expensive Olympics Ever. *The Christian Science Monitor*, p. 6.

Zygar, M., & Grib, N. (2009). Mister Neft. *Kommersant*, *78/P*(4133). Retrieved from http://kommersant.ru/doc.aspx?DocsID=1164267&NodesID=4

6 Conclusions and policy recommendations

Away from resource dependency toward a sustainable energy mix

Major trends

Focusing on oil and natural gas, as well as future renewable energy development, this book explores three important and somewhat puzzling bilateral cases of energy cooperation – U.S.–Russia, U.S.–Azerbaijan, and Russia–Germany – that vary in their geographic proximity, population and territory size, and power status, as well as their regime types. These cases represent major players in global energy security, with major emphasis on the United States and Russia. The recent conflict in Ukraine is also explored here, as it has had a profound impact on Russia–West relations.

Over the last two decades, there have been startling differences in the ways Russia has dealt with the United States and with Germany. Except for the brief period of the 'reset' in 2009–2010, the rest of the U.S.–Russia relations have been in a steady decline, reaching a new low almost every year. On the contrary, Russia has enjoyed seemingly stable, friendly relations with Germany. As about half of Russia's GDP has been oil and natural gas export revenues, it is not surprising that most of Russia's cooperation with Germany and the United States in the area of trade and investments have happened in the energy sector.

Other post-Soviet states have had much more friendly ties with the United States. One such state – Azerbaijan – has managed to maintain close economic and security ties with almost all of the neighborhood and beyond, and especially with the United States.

A structured comparison of U.S.–Russia and U.S.–Azerbaijan relations has revealed informative parallels and explanations. The main factor that has contributed to the tone of the bilateral relations has been the government's policy and the political will to utilize economic opportunities. Azerbaijan, being practically landlocked from Western Europe and the United States, managed to create a favorable business climate for international companies and other governments, inviting foreign investments and protecting them from domestic and external turbulence. Russia, on the contrary, has seen an opposition to the United States as an opportunity to strengthen the domestic regime. Rather than protecting foreign investors, Russia's business climate favored the Russian-government-controlled oil and gas companies. As a result, several American (and other Western) companies lost the business contracts that were available in the 1990s.

This distancing has not been the case with Germany, with which the Russian government favored and pushed multibillion-dollar projects, such as the Nord Stream pipeline. Both governments have been focused on extracting business opportunities from favorable geographic conditions and Russia's enormous oil and natural gas resources. Simultaneously, Germany has not been focused on Russia's domestic perturbations and the 'vertical of power'.

However, friendly Russian–German ties might change as the situation in Ukraine develops, after Crimea became a part of Russia, and Eastern Ukraine started a secession movement that led to a military conflict with thousands of casualties and millions of internally displaced people and refugees. The Western sanctions that followed the conflict will likely have a long-term disruptive effect on Russia's connections with the West.

It remains to be seen how the Ukraine–Russia conflict plays out in the energy security of Russia, Western Europe, and globally. The preliminary conclusion is that Ukraine and other countries have been switching to Norwegian natural gas and expanding the plans for LNG from the Middle East and even from the United States. Even though in 2015 the overall situation in the European energy mix has not yet changed drastically, the conflict has induced Western European countries to speed up the development of renewable energy. This development has been informally led by Germany, which has invested in renewables for several decades, despite having exceptional access to plentiful Russian natural gas.

All in all, the Ukraine conflict, while being a tragic event in itself, might bring a positive development the European economy, less dependent on hydrocarbon fuels. Such substitution for renewables will also be helping in global mitigation of climate change and global warming.

All three cases in this study, as well as the section on Ukraine, provide ample material for those interested in post-Soviet affairs and energy security. The chapters contain information about the economic development of Russia and Azerbaijan, their energy resources, major deals, and ups and downs in their relations with the United States and Germany. The chapter on Ukraine has been focused on the recent events: the February revolution in Kiev of 2014, the Crimea referendum in the same year, the military conflict in the Eastern Ukraine, and the Western sanctions and implications for energy contacts.

The sections below provide brief summaries of the three cases and the Ukraine conflict, as well as policy recommendations for the long-term and short-term.

The U.S.–Russia case

After the investment frenzy of the 1990s, American energy companies working in Russia experienced difficulties and setbacks in the 2000s. The frenzy began when Russia opened its economy and its energy sector to foreign investors shortly after the breakup of the Soviet Union. American businesses were encouraged by the Clinton administration to take advantage of opportunities in Russia, despite the economic turmoil, inflation, and overall instability ('Doing Business in the New Independent States – A Resource Guide', 1994). Regular meetings between Boris Yeltsin and Bill Clinton were amicable, and American advisers were invited to help with economic and political reforms (Lieven, 1998). In that period, Western energy companies – Chevron, ConocoPhillips, ExxonMobil, and others – started to work in Russia, signing production sharing agreements (PSAs) with the Russian government for such multibillion-dollar projects as Sakhalin-1 ('Russia-Sakhalin-1', 2010).

In the 2000s, driven by oil and gas revenues and soaring oil prices, the incoming Putin administration was able to strengthen its domestic and international position. As a result, Russia became more assertive on such issues as NATO's expansion, U.S. missile defense in Europe, and the Iranian nuclear program (Goldman, 2008). Simultaneously, the share of American investments in Russia drastically dropped, while other countries maintained their investment levels. Russia tightened government controls over domestic oil and natural gas resources that received the status of the highest strategic importance for the country's economic security. As a result, major Western energy companies did not sign new large contracts in Russia for most of the decade, and the Russian government renegotiated the existing PSAs on environmental and tax grounds (Donohue & Shokhin, 2008). In 2008, during the Obama–Medvedev 'reset', opportunities for major energy companies started to reappear, mostly in the Arctic and the remote East Siberian oil and natural gas fields (Cohen, 2011).

For a while, before the 'shale revolution', the United States considered possibilities of importing liquefied natural gas (LNG) from other countries, including Russia. In addition to political obstacles and geographic challenges, another reason why American companies seem no longer interested in importing Russian LNG is the soaring domestic production of shale gas in the United States that has changed the domestic supply-demand balance for natural gas. Since 2011, the United States has been self-sufficient in production of natural gas and does not need expensive imports. This change occurred extremely quickly and unexpectedly: as recently as in 2009, North America was Gazprom's primary destination for liquified natural gas ('Gazprom to Sell as Much as 90% of Shtokman LNG in North America', 2009). By 2011, the United States drastically increased shale gas production, which drove domestic prices down, making Russian LNG imports to the United States unfeasible. As a result, Russia diverted its LNG exports to Asia.

U.S.–Russia energy ties still experience difficulties because of strong intergovernmental disagreements. In this case, the geopolitical explanation prevails, especially after Vladimir Putin's return to the presidency in 2012 and the conflict in Ukraine in 2014, when U.S.–Russia relations deteriorated again, and the future of new energy deals is in doubt. During Putin's third term, Russian energy lobbies have been even more intertwined with the government, as energy executives simultaneously hold government positions and promote government agendas.

The U.S.–Azerbaijan case

U.S.–Azerbaijan is an interesting case of growing energy cooperation despite such seemingly unfavorable conditions as Azerbaijan's distant geographic position with no ocean access and a strong Armenian lobby in the United States. Additionally, estimates of Azerbaijan's oil reserves differ widely in different sources – from 7 to 233 billion barrels (Blanchard, 2005) – but nevertheless the country is one of the key energy partners for the United States. The United States strongly supported the Baku–Tbilisi–Ceyhan (BTC) pipeline that now delivers oil from Azerbaijan to Turkey via Georgia.

The BTC pipeline was successfully built by a consortium of Western companies, despite strong opposition by the Armenian lobby in the United States, which has been powerful in multiple intergovernmental issues. It was successful in blocking other government decisions, such as the recall of U.S. Ambassador to Azerbaijan Matthew Bryza (who was accused of pursuing Azerbaijan's interests) and blocking the Turkey–Armenia rapprochement (Abbasov, 2012). Based on these successful instances, one might expect that the Armenian lobby would be able to block the American support of the BTC pipeline or to divert the pipeline's route to Armenia instead of Georgia. However, this pipeline was of the highest importance for the U.S. government, which was implementing the policy of multiple pipelines in order to reduce Russia's influence in the region (Alieva, 2011). As a result, the U.S. government backed the pipeline route and ensured that it was successfully built.

Over the last two decades, Azerbaijan has substantially improved its economic climate that created favorable conditions for foreign investors. However, the economic motives alone are not sufficient to explain why the United States and other Western governments supported the BTC route, because of the route's costs and technological challenges. The BTC route was selected from several alternatives: other proposed routes would have gone via Russia, via Iran, or via Armenia ('Armenian Lobby Applies Pressure to Get BTC Passing Through Armenia', 2001). Among these options, the BTC route was considered by experts as the most expensive and technologically challenging one (Shaffer, 2009), complicated by Georgia's secessionist conflict in South Ossetia and Abkhazia, which could potentially endanger the pipeline. To a large extent, the BTC route was selected on geopolitical grounds. Since the early 1990s, the government of Azerbaijan has maintained close relations with the United States and its allies.

The Russia–Germany case

Russia–Germany energy cooperation is an unusual case in which the two countries have overcome their past Cold War enmity and pursued joint projects despite the opposition from the EU and the United States. One of such projects, the $10.2 billion Nord Stream natural gas pipeline, which was built on the bottom of the Baltic Sea and bypassed European transit states, was surprising to many, taking into account that the European Union had been working on energy source diversification in order to reduce its dependency on Russia (Steininger, 2011). The pipeline boosted Germany's position as the European energy hub and reduced transit states' (Poland and the Balkans) revenues and influence in the European Union. As a result, Nord Stream caused an outcry in multiple circles – new EU member-states, environmental NGOs, and Western think tanks (Kramer, 2009).

The proximity between the two countries and Russia's vast energy resources ensure a powerful economic incentive and make the oil and natural gas trade between Russia and Germany economically justifiable. Nevertheless, it took the political will of the two governments who actively promoted their cooperation in the energy sector, in which Germany and Russia show a strong case of mutual dependency. Since the Cold War era, Russia has admitted that it needs German technologies in exchange for Russia's raw materials, oil and gas. Multibillion-dollar projects (such as the Nord Stream) have materialized only as a result of governments' decisions, rather than via 'pure' market mechanisms.

Nord Stream helped strengthen Germany's position as a European energy hub and natural gas distributor, providing Germany with direct access to Russia's oil and gas. Simultaneously, Russia has kept its near-monopolistic role in European energy supplies. The domestic business lobbies in both countries (e.g., associations of energy businesses and European business associations) were closely aligned with their governments and intensively advocated the pipeline. Other societal groups, such as environmental NGOs, were opposed to the pipeline but did not have enough influence to block the Nord Stream project.

Connection with international relations theories

The research above started with a question: why are some international energy projects, such as pipelines or joint productions of oil and natural gas, actively implemented while others, seemingly equally feasible and possibly even more profitable, fail to materialize? Why are some states willing to promote bilateral energy trade and investment while others seem unable to take advantage of mutually beneficial energy projects? To answer these questions, this study has scrutinized three important post-Cold War interstate energy relationships to evaluate the relative importance of *power, profits, and politics*.

More specifically, relevant theories of international relations – realism, liberalism, and domestic interest groups – have been applied here to explain interactions in the energy security area. Realism (Krasner, Gilpin, and Socor, etc) stresses the prevalence of states' considerations of geopolitical survival, power, relative gains, and zero-sum thinking. Examples include the legacy of the Cold War, military alliances, and security dilemmas. According to realism, a government will trade in energy resources only if such trade and investment are in its national interest, in which security concerns prevail over profit maximization.

Liberalism (Keohane, Nye, Copeland, Sachs, Banks, Stiglitz, etc.), on the contrary, stresses the importance of absolute gains and argues that energy interdependence is a peace-inducing factor: states are able to trade and cooperate even when there are political and ideological disagreements. Liberalism in energy cooperation focuses on the economic fundamentals of projects – the availability of oil and gas resources, the demand for them, and the investment climate for joint oil and gas production, as well as geographical opportunities for export and import.

Finally, the literature on domestic interest groups, from the classic works of the 1960s to contemporary studies (from Baumgartner, Leech, Foyle, and Belle to Franklyn Griffiths, and Pavel Tolstykh, etc) provides an insight on lobbying and other activities of influential business, ethnic, and other societal groups. Such interest groups promote their political and economic goals and may advance or oppose specific policies in the energy sector.

Each of these theoretical approaches provides valuable knowledge that sheds light on various aspects of energy cooperation, but alone each is narrow and one-sided, unable to embrace the broad complexity of real-world energy relations. This study offers a unified framework that allows simultaneous testing of their theories and a deeper understanding of important on-going cases in energy relations.

The unified framework differs from existing energy security explanations in that they typically 'test' only a single theory or simply employ a descriptive historical approach. While historical work gives us extremely useful knowledge of how events have unfolded in the energy domain, the framework developed in this book will help offer a more generalizable explanation of energy trade and investment outcomes.

It has been clear in the explored cases that geopolitical rivalries are extremely influential in determining energy relations between states, but economic considerations can help overcome geopolitical rivalries. In some cases, such as Nord Stream, despite previous geopolitical or historical enmity, governments choose to use their political wills to overcome these differences for the sake of mutually profitable energy projects. Finally, domestic interest groups may be influential in promoting their common interests, if they are united and organized around an energy issue. However, oftentimes, powerful lobbies divide their attention and prioritize other, non-energy issues, which again allow the government line to dominate in decision-making. The cases explored and tested here provide empirical support for these developments.

Long-term lessons from the cases

Studies such as this book provide empirical insight in energy security literature and lessons for policy-making. The three cases – U.S.–Russia, U.S.–Azerbaijan, and Russia–Germany – have provided several lessons that can be applied to other cases.

The first lesson is that geopolitical rivalries, even in non-energy spheres, are the main obstacle to mutually lucrative energy deals. Therefore, it is important to take into account the issue-linkage between energy negotiations and non-related security concerns, such as nuclear weapons, war, human rights, and nationalism.

For example, Russia's concerns about NATO's expansion and other military issues affect Russia's decisions about oil and gas production deals, awarding these contracts to 'friendly' governments. Russian Ambassador to NATO Dmitry Rogozin complained in 2010 about 'a unilateral world, NATO-centrism, the alliance's spontaneous expansion eastwards and refusal to recognize the principle of integrity and security' (Moran, 2010). Since energy projects in Russia are very closely controlled by the Russian government, any political disagreements between governments can negatively affect the outcome of energy negotiations, such as those between the United States and Russia.

The West's position in negotiations with Russia has been somewhat helped by the recent global economic recession, which reduced the world demand for Russian energy resources. The reduced demand, in turn, has weakened Russia's bargaining position on European energy markets:

> Recent renegotiations of long-term gas price contracts with the Russian monopolist by major European utilities (including Germany's E.On AG, Austria's OMV, and Italy's Edison and Eni, France's GDF Suez (GSZ) SA) [have] showed Gazprom's willingness to give concessions. Such leverage on Moscow may weaken in a few years, as gas demand picks up [the] pace… in Europe.
>
> (Berdikeeva, 2012)

The second lesson is that countries could start treating competing projects (such as two different pipelines) not as a zero-sum game but as an array of opportunities favorable for all the involved parties. Following the liberal agenda of pursuing absolute gains and creating projects of mutual advantage may help overcome the 'us versus them' mentality, which has previously deadlocked multiple negotiations.

When governments seem to be unable to reach a compromise, energy companies step in and offer solutions. For example, in 2010, to overcome the stalemate in the Nabucco and South Stream pipeline discussions, the Italian energy company ENI suggested a different way of transporting natural gas in the Caspian region. ENI suggested shipping natural gas by sea in compressed form from Turkmenistan to Azerbaijan, eliminating the need for a pipeline (Socor, 2012). This solution would bring natural gas to consumers without constructing either pipeline, yet still reaching the goal of energy security for the region.

The third lesson is a possibility to delegate decision-making in the energy domain to international organizations, such as the International Energy Agency and the World Bank's energy division, which could provide a more objective insight about future energy routes. If decision-making could be at least partially delegated from governments to other levels – international, industry, and project – it would be possible to achieve more absolute gains for all involved countries (Razavi, 2007). Independent experts in international institutions could evaluate various options and offer policy recommendations. Thus, decision-making would be less politicized, allowing the selection of the most economically, technologically, and financially efficient projects.

Understandably, such solutions might not be very realistic in the near term, since governments are unlikely to cede their power over energy resources and energy deals. In this case, short-term energy solutions might be more applicable.

Short-term policy recommendations

The above lessons contain implicit policy recommendations, although they might be too idealistic to be implemented because of strong government control in the energy sector. However, there is another realm of energy cooperation where tensions are lower, projects are more modest, and political statements are less complicated. So, starting at this level, small-scale projects could eventually lead to greater cooperation in projects of high strategic importance. The sum-total of these small projects could establish greater trust, which could help to overcome impasse in larger investment projects.

Collaboration could start in areas less sensitive geopolitically such as renewable energy, environmental protection, energy conservation, refining, and oil-drilling safety. These are the areas in which the United States and Germany have vast knowledge, which the former Soviet states lack. More cooperation in these aspects would create a better business environment for future joint projects in the strategically important sector of oil and gas production. Promoting these less sensitive projects would help improve the economic climate for subsequent large-scale oil and gas deals.

The international relations literature has long addressed analogous cases of this type of trust-building cooperation, in which the reduction of tensions was very gradual and thus less inflammatory. Cooperation starts small and gradually leads to larger agreements. One example is nuclear negotiations and disarmament between the USSR and the United States, which started with modest reductions and led to major agreements such as START-1 and START-II, among others. This is a typical example of the gradual reduction in tension (GRIT), theorized by Charles Osgood as 'de-escalation by making a small, unilateral (one-sided) concession to the other side, and at the same time, communicating a desire or even an expectation that this gesture will be matched with an equal response from the opponent' (Osgood, 1962). Similarly, in the area of political economy, such scholars as Peter Evans in his 'Is an Alternative Globalization Possible?' (2008) and Peter Hall in *The Political Power of Economic Ideas* (1989) explore how gradual concessions and small steps in economic liberalization lead to larger-scale economic cooperation and trade agreements among states.

Another instance of incremental policies becoming bigger milestones was the abolition of South African apartheid via informal activist networks (Keck & Sikkink, 1998). The anti-apartheid movement was formed at smaller conferences and included incremental legal actions, leading to United Nations laws that became legally binding and imposed harsh, punitive sanctions. These gradually increasing actions put a stop to this abhorrent human practice and culminated in an international regime backed by the United Nations.

In the energy domain, less contentious, non-military projects (e.g., climate change research, carbon emission reduction, energy efficiency, and renewable energy) can be implemented with government backing and federal support. These projects might be especially beneficial for oil-dependent countries (Russia, Azerbaijan) to break out of their resource-income dependency and to become competitive on the world market (Desai & Goldberg, 2008). Investing some current oil revenues in 'green' technologies would help oil-dependent states diversify their economies. An example of such programs is the 'Smart Energy' initiative by Green Cross Russia. This program unites scientists, environmental activists, and politicians in shaping a strategy of alternative energy development. The program covers all kinds of alternative energy – wind, geothermal, solar, and biomass (Chumakov, 2008).

Renewable energy could be an area in which scientists, NGOs, and governments productively collaborate. In the area of green energy, Russian scientists already do advanced research on wind turbines, photoelectric systems, lumber production waste, solar water heating systems, bio energy from manure and harvest waste, anaerobic fermentation of agricultural and forestry waste, solar energy, and municipal waste burning. NGOs, which are the main agenda-setters for alternative energy development, could bring such scientists together, with continued government support to diversify the energy mix (English, 2000).

Government support of green energy started in Russia in the 2000s, when the Russian government funded research and development in renewable energy. It also implemented 'green certificates' – a program similar to the Western cap and trade system. Via tax incentives and special prices for renewable energy, the government designed policies for encouraging 'green' development (Zerchaninova, 2006, p. 455). After a while, however, oil and natural gas again became the priority, as oil prices were record-high in the mid-2000s.

In 2012, the Russian government again attempted to reform the energy sector and to improve energy efficiency using modern technology. According to the director of the Center for Efficient Energy Use, Igor Bashmakov, the government's goal has been to reduce energy use by 40 per cent by 2020, cooperating with scientists and independent energy experts. However, these goals have been slowed down by bureaucratic red tape and corruption, including an ineffective system of energy audits and energy passports (Davydova, 2012). Cooperation with Western NGOs and governments could help Russia reduce its bureaucracy and paperwork, bringing Western business practices and cutting-edge research to outdated Russian enterprises.

Russia has already cooperated with the United States on energy efficiency. The first joint research projects and publications started back in 1983 between the Soviet Academy of Sciences and Princeton's Center for Energy and Environmental Studies (and later the U.S. National Academy of Sciences) after the scientists on both sides had realized 'that the stifled communications between the United States and the USSR had blinded both sides to the other's energy-technology advances' (Millhone, 2010, p. 15).

In the newly independent Russia of the 1990s, the U.S. government opened energy efficiency centers and included the country in the Country Studies Program on climate change. The first of these centers, the Russian Center for Energy Efficiency (www.cenef.ru), was jointly sponsored by the U.S. government, the World Wildlife Fund, and the International Social and Ecological Union (Russia). As Russia's leading think tank for energy efficiency research, it has provided policy recommendations for energy efficiency improvement, cooperating with governments and major international foundations: the U.S. Agency for International Development, Department of Energy, Environmental Protection Agency, the World Bank, United Nations Environment Program, International Finance Corporation, Global Environment Facility of the United Nations Development Program, European Bank for Reconstruction and Development, United Nations Economic Commission for Europe, The John D. and Catherine T. MacArthur Foundation, Alliance to Save Energy, Natural Resources Defense Council, and the International Institute for Energy Conservation.

Energy efficiency discussions have been closely tied to climate change research. In the early 1990s, the Russian Federal Service for Hydrometeorology and Environmental Monitoring (RosHydroMet) started working with the U.S. Country Study Program on GHG emission reduction, followed by the U.S. Initiative on Joint Implementation and Clean Development Measure within the Kyoto Protocol regarding heating, reforestation, and fugitive gas capture (Millhone, 2010, p. 16).

After a low level of interest in the last decade, when the Russian government was more concerned with economic growth, the topic of energy efficiency has recently been revived by then President Dmitry Medvedev's efforts to modernize the Russian economy, admitting that the country's overall energy efficiency is very low and leads to enormous energy losses. John Millhone of the Carnegie Endowment for International Peace compared these losses to 'untapped energy wealth' (Millhone, 2010).

Major energy companies, such as those that invest in Russia, have also joined the renewable energy race. Oil companies, previously skeptical, have launched multiple alternative energy projects. For example, in 2009, ExxonMobil announced that it would invest heavily in algae and other biofuels (up to $600 million per year). Biofuel production has also enjoyed support and incentives by the U.S. government, which has set the target of 36 billion gallons of biofuels produced in the United States per year. Currently, biofuels (mostly corn) already provide 9 per cent of U.S. fuel consumption ('ExxonMobil Will Invest in Algae to Fuel Development', 2009).

ExxonMobil, BP, and other energy companies are interested in investing in alternative energy projects abroad, including post-Soviet states. Such joint ventures could bring American advanced technology and capital to Russia, Azerbaijan, and Kazakhstan, while utilizing scientific talent in chemistry and biology. In addition to biofuels, such projects could be started in battery production, as well as oil shale, ethanol, and liquid coal production (Blanchard, 2005, p. 266). These joint projects in 'green' energy could be a potentially lucrative area with little geopolitical contention, providing high-paying jobs for local scientists.

Russia is currently the third largest emitter of greenhouse gases in the world, and the country's energy consumption could be reduced by half via more efficient technology. Greater energy efficiency and renewable energy sources (e.g., wind) would also help Russia solve the problem of electricity blackouts in the Russian Far East (Newell, 2004, p. 62).

One more renewable resource, plentiful in Russia, is wind. In the area of wind energy, the U.S. National Renewable Energy Laboratory (NREL) have already installed wind-diesel hybrid systems in Northern Russia ('Wind Energy Applications Guide 2001', 2001). Examples of wind turbine importers in Russia include the Global Wind Energy Council and Bergey Windpower (USA) ('Global Wind Energy Market Report: Wind Energy Expands Steadily in 2004', 2005). With the Russian government's encouragement, more such American companies could invest in Russia's wind energy, diversifying Russia's energy mix.

Unless Russia, Azerbaijan, and other post-Soviet oil-dependent states modernize and diversify their economies and reduce their dependence on petroleum revenues, the governments have little choice but to pump out more oil and gas. Any reduction in global petroleum demand or a drop in oil prices would be catastrophic for oil-dependent economies. Cooperating with foreign companies and governments in 'green' technologies would help Russia and Azerbaijan break out of the 'resource curse' (Ross, 2012).

At the same time, joint 'green' projects would increase the population's exposure to Western values and practices in high-tech industries, potentially facilitating societal change in Russia. Cooperation in projects of low contention would also help alleviate Cold War fears and mentality. As a result, one would likely see the creation of a possibly more democratic and economically sound Russia, which would be a better partner for the West, both in energy projects and beyond.

In sum, energy development and cooperation are of obvious critical economic importance. Yet as with so many other international relations issues, considerations of power often stand in the way of profits (not just economic, but political-diplomatic as well). This is what the cases show, and the conclusions do not point to significant energy cooperation progress in the future if the status quo continues. However, cases from other arenas of international cooperation (arms control, trade negotiations, human rights) show how incremental policies – which begin small – come to influence larger policies; peripheral policies become core, quantitative progress begets qualitative change, ultimately allowing mutual *profit* to overcome the limitations of *power*.

References

Abbasov, S. (2012, January 5). Azerbaijan: Baku Fumes Over Scuttled Ambassadorial Appointment. *Aurasianet.Org*. Retrieved from http://www.eurasianet.org/node/64796

Alieva, L. (2011). Regional Cooperation in the South Caucasus: The Azerbaijan Perspective. In M. Aydin (Ed.), *Non-Traditional Security Threats and Regional Cooperation in the Southern Caucasus* (pp. 197–207). Amsterdam: IOS Press.

Armenian Lobby Applies Pressure to Get BTC Passing Through Armenia. (2001, July 28). *Hurriyet Daily News*. Retrieved from http://www.hurriyetdailynews.com/default.aspx?pageid=438&n=armenian-lobby-applies-pressure-to-get-btc-passing-through-armenia-2001-07-28

Berdikeeva, S. (2012, May 10). Taking a Second Look at the Southern Gas Corridor. *OilPrice.Com*.

Blanchard, R. D. (2005). *The Future of Global Oil Production: Facts, Figures, Trends and Projections, by Region*. Jefferson, NC: McFarland & Company, Inc.

Chumakov, A. (2008). Green Cross Russia: 'Smart Energy'. *Green-Cross.Ru*. Retrieved from http://www.green-cross.ru/programms/energy/

Cohen, S. (2011, June 20). Obama's Russia 'Reset': Another Lost Opportunity. *The Nation*.

Davydova, A. (2012, September 28). They Had to Return to the Issue of Lightbulbs. *Kommersant*. Retrieved from http://kommersant.ru/doc/2032143

Desai, R. M., & Goldberg, I. (2008). Introduction. In R. M. Desai & I. Goldberg (Eds.), *Can Russia Compete?* (pp. 1–11). Washington, DC: The Brookings Institution Press.

Doing Business in the New Independent States – A Resource Guide. (1994) (July 1994 ed., Vol. 5, pp. 497–511). U.S. Department of State Dispatch.

Donohue, T., & Shokhin, A. (2008). U.S.-Russia Business Dialogue. *RIA Novosti*, (July 28, 2008). Retrieved from http://rbth.ru/articles/2008/07/28/Business_dialogue.html

English, R. (2000). *Russia and the Idea of the West: Gorbachev, Intellectuals, and the End of the Cold War*. New York, NY: Columbia University Press.

Evans, P. (2008). Is an Alternative Globalization Possible? *Politics & Society*, *36*(2), 271–305.

ExxonMobil Will Invest in Algae to Fuel Development. (2009). *Oil and Gas Eurasia*. Retrieved from http://www.oilandgaseurasia.com/news/p/0/news/5254#

Gazprom to Sell as Much as 90% of Shtokman LNG in North America. (2009). *Oil and Gas Eurasia*. Retrieved from http://www.oilandgaseurasia.com/news/p/0/news/5923#

Global Wind Energy Market Report: Wind Energy Expands Steadily in 2004. (2005). Retrieved April 4, 2009, from http://www.awea.org/pubs/documents/globalmarket2005.pdf

Goldman, M. (2008). *Petrostate: Putin, Power, and the New Russia*. Oxford: Oxford University Press.

Hall, P. A. (1989). *The Political Power of Economic Ideas: Keynesianism across Nations*. Princeton: Princeton University Press.

Keck, M. E. and Sikkink, K. (1998). *Activists beyond Borders*. Ithaca, NY: Cornell University Press.

Kramer, A. E. (2009, October 12). Russia Gas Pipeline Heightens East Europe's Fears. *New York Times*. Retrieved from http://www.nytimes.com/2009/10/13/world/europe/13pipes.html

Lieven, A. (1998). *Chechnya: Tombstone of Russian Power*. New Haven: Yale University Press.

Millhone, J. P. (2010). Russia's Neglected Energy Reserves (pp. 1–54): Carnegie Endowment for International Peace.

Moran, A. (2010, February 25). Russia: Russia will Respond to NATO Expansion, U.S. Missiles. *Digital Journal*.

Newell, J. (2004). *The Russian Far East: A Reference Guide for Conservation and Development*. McKinleyville, CA: Daniel & Daniel.

Osgood, C. E. (1962). Studies on the generality of affective meaning systems. *American Psychologist*, 17(1), January 1962, 10–28.

Razavi, H. (2007). *Financing Energy Projects in Developing Countries*. Tulsa, OK: Penn Well Corporation.

Ross, M. L. (2012). *The Oil Curse: How Petroleum Wealth Shapes the Development of Nations*. Princeton, NJ: Princeton University Press.

Russia-Sakhalin-1. (2010). ExxonMobil.

Shaffer, B. (2009). Permanent Factors in Azerbaijan's Foreign Policy. In A. Petersen & F. Ismailzade (Eds.), *Azerbaijan in Global Politics: Crafting Foreign Policy* (pp. 67–84). Baku, Azerbaijan: Azerbaijan Diplomatic Academy.

Socor, V. (2012). Italy Ambivalent About Gazprom's South Stream Project. *Eurasia Daily Monitor*, 9(142).

Steininger, M. (2011). Nord Stream Pipeline Opens, Russia-Europe Interdependence Grows. *The Christian Science Monitor*. Retrieved from http://www.csmonitor.com/World/Europe/2011/1108/Nord-Stream-pipeline-opens-Russia-Europe-interdependence-grows

Wind Energy Applications Guide 2001. (2001). Retrieved April 8, 2009, from http://www.awea.org/pubs/documents/appguideformatWeb.pdf

Zerchaninova, I. L. (2006). Vozobnovlyaemaya Energetika: Globalnie Tendentsii i Stimuli Razvitiya. In A. T. Nikitin & S. A. Stepanov (Eds.), *Globalnie Problemi Bezopasnosti Sovremennoi Energetiki* (pp. 444–459). Moscow: Publishing House of IIUEP.

Index

For Product Safety Concerns and Information please contact our EU
representative GPSR@taylorandfrancis.com
Taylor & Francis Verlag GmbH, Kaufingerstraße 24, 80331 München, Germany

www.ingramcontent.com/pod-product-compliance
Ingram Content Group UK Ltd.
Pitfield, Milton Keynes, MK11 3LW, UK
UKHW020954180425
457613UK00019B/680

* 9 7 8 0 3 6 7 3 3 2 7 3 0 *